**The Future of
Educational
Telecommunication**

The Future of Educational Telecommunication

A Planning Study

George W. Tressel
Donald P. Buckelew
John T. Suchy
Patricia L. Brown
Battelle Memorial Institute
Columbus Laboratories

Lexington Books
D.C. Heath and Company
Lexington, Massachusetts
Toronto   London

Library of Congress Cataloging in Publication Data

Main entry under title:

The Future of educational telecommunication.

Bibliography: p.
1. Educational broadcasting—United States. 2. Educational broad-
casting—Law and legislation—United States. 3. Telecommunication—
United States. I. Tressel, George W.
LB1044.8.F87   384.54   74-27645
ISBN 0-669-97691-1

Published simultaneously in Canada

Printed in the United States of America

International Standard Book Number: 0-669-97691-1

Library of Congress Catalog Card Number: 74-27645

Contents

List of Figures

List of Tables

Foreword

The goals of educational broadcasting are broad and ambitious, embracing both instructional and cultural activities, but modified by the limited capabilities of an elaborate and costly distribution network. Thus, the Educational Broadcasting Facilities Program stimulates these activities, but to some extent it also modifies their nature because of funding limitations. The field has many facets, responds to a variety of needs and pressures, and competes for audience with a well funded industry whose success is measured in far simpler terms.

To better meet the challenge of planning for the future, the U.S. Office of Education asked the Communication Research Laboratory of Battelle-Columbus to survey the goals, status, and future of educational broadcasting and make recommendations based upon this perspective. The primary source of information has been a review of the extensive available literature. We have organized the facts and synthesized opinions from a large number of publications, tempering the material with our own opinions and those of many other knowledgeable members of the field. We visited a number of facilities and talked to over 60 managers of educational broadcast facilities. They represent both radio and television stations, large and small, from all parts of the country; and were chosen to represent a variety of ownership patterns. We also visited managers of the major agencies and foundations which are concerned with funding, policy, management and control; collecting both factual information and experienced opinion on the goals and problems of the industry.

After summarizing the goals of educational broadcasting as articulated in the laws and industry opinion, we compared them with our best information on the status, health, accomplishments, and inadequacies of the system. It is not possible to present a simple picture . . . educational broadcasting is a melange of activity in pursuit of two ephemeral goals: improved education and a cultural alternative to commercial broadcasting. Style and viewpoint vary widely— ranging from austere intellectual and cultural innovator to entrepreneur and hardware collector.

In scanning the future we drew heavily on several of our previous studies of the state of technology, prepared for the U.S. Army Communications Electronics Directorate. The effect of this emerging technology will probably be to accelerate and emphasize current trends and problems rather than to produce fundamental change.

Our recommendations emphasize the critical need for "full coverage" and competitive technical quality, i.e., a basic distribution system comparable to that of commercial broadcasting. In view of limited resources we recommend that three levels of production capability be defined and resources allocated to provide equitable geographical distribution of this capability. At the same time we believe that current guidelines must be broadened to accommodate new technology and methods of instruction. We expect that instructional and public broadcast activities will soon follow distinctly different growth patterns.

Educational broadcasters we contacted in making this study were most cooperative. On short notice, many of them interrupted busy schedules to provide hours of time, answering our questions with frankness and candor. Staff members of the U.S. Office of Education were most helpful, particularly Mr. Arthur S. Kirschenbaum of the Office of Planning, Budgeting, and Evaluation, who served as Project Officer. Mr. Kirschenbaum made many helpful comments, but in no case did he make changes which influenced the outcome of the Battelle study.

Mr. Donald Patrick contributed the illustrations which enliven several pages of this book and provided valuable advice on the graphics of report preparation. The original manuscript was typed by Mrs. Lynette Parmer and Mrs. Janice Jarvis, with skill and with care.

Preface

Educational telecommunications are in a state of flux that is part of the entire telecommunications industry. In perhaps no other area of technology are changes coming so fast. A communications satellite is launched, and new delivery systems are tested in Appalachia, Alaska, and the Rocky Mountains. . . . . Far-reaching decisions are being made by the courts and the Federal Communications Commission on the roles of common carriers. . . . Domestic (as well as international) commercial satellite systems are entering the picture. . . . The impact of new technology is being felt as slow-scan television, interactive on-line computer systems, broadband transmission, and other innovations are opening up enticing possibilities.

The impact of federal policies adds to the state of flux. As cable television begins to assume the proportions of an alternative to over-the-air broadcasting, regulations are promulgated to protect both public and private interests. As inflation becomes a pervasive influence in every economic area, larger budgets, at last, are granted educational telecommunications practitioners. A new national matching-fund policy is formed which, hopefully, will provide educational broadcasters with $100 million in federal funds per year by 1979. Greater local autonomy is advocated as part of this policy.

In such an environment it is useful to "freeze the action" at finite points in time to provide insight, thoughtful consideration, and mileposts against which to judge future change. It is our intent to provide such a milepost as of late 1973. It also is our aim to develop a fresh, objective viewpoint that reflects knowledge of the policies and problems of educational broadcasting but is not attenuated by concern with problems of operation, however important they are, that do not directly affect policy formation.

This report illustrates the impact of federal policies related to educational broadcasting, but it does not attempt to judge these policies. Born in a world of give-and-take, educational broadcasting necessarily is affected by compromises, which are part of the American political process. We are not gifted with second sight to project what might have been had there been different compromises. Rather, our purpose must be to develop a logic for the future of educational telecommunications based upon what has actually happened.

Throughout this report, the terms *educational broadcasting, instructional broadcasting,* and *public broadcasting* are used in three different contexts. Educational broadcasting is the entire service rendered by "educational television," "educational radio," and other educational telecommunication forms. The "public" and "instructional" parts of educational broadcasting have developed in parallel, but such development is not absolutely necessary, because the goals are not the same. Instructional broadcasting is the component directly related to supplying formal instruction at any level. Public broadcasting includes public affairs, cultural, and entertainment components. We do not use the term *noncommercial broadcasting,* because this means any program or offering that does not have a commercial sponsor, including religious programs, network public affairs, etc. Noncommercial broadcasting in other countries (for example, the BBC) may be a general service intended for mass audiences, rather than a service specifically intended to provide educational or cultural enrichment.

1 Introduction

The Face of Educational Broadcasting

In 1973, eleven years after the first federal funding and six years after the Public Broadcasting Act of 1967, educational radio and television began to mature and develop a character of their own. Despite the fact that only one out of four adults watched public television (and then for less than two hours per week), programs such as "Civilization" and "Sesame Street" achieved national recognition as well as artistic success.

A variety of agents have entered the field; radio stations are generally operated by colleges and universities, but television stations are divided among educational agencies, state agencies, and nonprofit community corporations. In this first decade a national television distribution system has been established as well as six regional systems and at least twenty-eight state networks. Seven major "production centers" and more than thirty other stations produce material for national distribution.

Despite program budgets that average one-fourth that of their commercial counterparts, public broadcasts periodically tweak the audience polls, offend politicians, and otherwise show signs of significant impact. This growing strength, however, provokes questions of priority and purpose. Is popular appeal a measure of artistic quality? Will education via entertainment produce a generation without patience for the tempo of the real world? As success (and income) are increasingly measured by audience reaction, will not educational broadcasting tend to become a copy rather than an alternative to commercial broadcasting?

The Educational Broadcasting Facilities Program (EBFP) has been the major stimulating force, though its dollars represent only a minor part of the capital investment in this industry. While this investment is small compared to operating budgets, it plays a significant role in the health and growth pattern of educational broadcasting. The EBFP is administered by the U.S. Office of Education, perhaps as much by historical accident as by design, and its goals and philosophy have never been sharply defined. This reflects the difficulty of reducing broad intangible goals to practical policy, rather than a lack of concern. It is for this reason that Battelle was assigned the task of surveying the status of educational broadcasting, weighing this against its goals, and considering ways in which the EBFP might help to meet these goals.

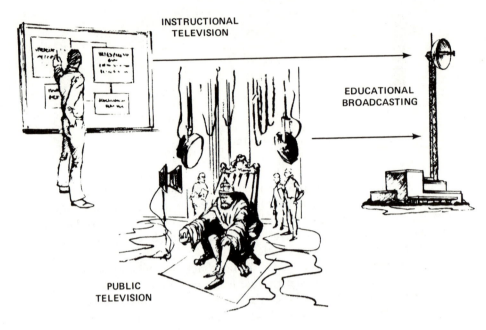

Figure 1-1. Distribution and Programming Pattern.

Broadcasting is a complex and costly medium. Issues of hardware often affect (and sometimes overshadow) the system they are meant to support. So it is not possible to isolate such questions, and we will attempt to consider them within the context of their instructional, cultural, and community service goals.

Educational broadcasting began as the exploration of a potential new medium for *instruction*. Later, as public, foundation, and government representatives voiced a need for *cultural alternatives* to the commercial "wasteland," educational broadcasting was assigned this additional role. The two roles are complementary in many ways, but we will examine the nature of this symbiosis in detail in the following pages.

Both broadcast functions require heavy investment in program production and distribution equipment, as well as in supporting technical and production personnel. To a great extent it is optimization of this investment resource that dictates the dual marriages of program production and transmission; instructional broadcasting and public broadcasting (figure 1-1). A number of ambiguities and inconsistencies accompany efforts to reconcile this ambivalence. It is our hope to clarify these problems rather than to provide a rationale for them.

It was not the intent of this study to collect *new* data. Our assessments are based on the evaluation of published information, tempered by firsthand impressions and conversations with knowledgeable members of the industry. We visited nine facilities and conducted over sixty in-depth interviews with educational broadcast licensees. (See Appendix A.) We also visited policy-makers (see Appendix B) in the principal agencies concerned with educational broadcasting, including:

1. Corporation for Public Broadcasting (CPB)
2. Public Broadcasting Service (PBS)
3. Ford Foundation
4. National Public Radio (NPR)
5. National Association of Educational Broadcasters (NAEB)
6. Federal Communications Commission (FCC)
7. Office of Telecommunications Policy (OTP).

An evaluation of educational broadcasting must recognize the possible effect of emerging technologies. The study considers those technologies that might be integrated into the existing system or represent potential alternatives for programming and distribution.

We have reviewed a variety of existing surveys, studies, public records, and other published data as well as a number of unpublished documents. We are deeply indebted to the many persons who contributed such material as well as their thoughts and opinions.

We believe that this report presents a reasonably objective synthesis of the facts, ideas, and opinions we gathered, and we hope that it provides useful planning assistance. Yet the utility of *any* report on educational broadcasting must be tempered by the resources that are available to the industry. The situation can be summed up in the fact that in a country where we spend almost $14 per capita each year for commercial television, we spend less than 70 cents for public television, only 20 cents of which is federal money.

2 The Goals of Educational Broadcasting

The goals of educational broadcasting are necessarily broad, embracing the initial concern for improved education as well as the subsequent quest for broadcasting alternatives and an expanded cultural and informational environment. A complete discussion of these goals must consider what is written in the applicable public laws and how policy-makers interpret the laws.

Educational Television Facilities Act of 1962

The first direct federal support of public broadcasting began in May 1962, when the Communications Act of 1934 was amended. This amended Act (Public Law 87–447) became known as the Educational Television Facilities Act of 1962. Its stated purpose was:

> to assist (through matching grants) in the construction of educational television broadcasting facilities . . . to achieve
>
> 1. *Prompt and effective use* of all educational television channels remaining available,
> 2. *Equitable geographical distribution* of educational television broadcasting facilities throughout the states, and
> 3. Provision of educational television broadcasting facilities which will serve the *greatest number of persons* and serve them *in as many areas as possible,* and which are adaptable to the *broadest educational uses.* [1] (Italics ours.)

The Carnegie Commission

In 1964, at a conference held by the National Association of Educational Broadcasters and with the cooperation of the U.S. Office of Education, Mr. Ralph Lowell proposed the establishment of a commission to analyze the financial needs of educational television and suggest how funding requirements might be met. This proposal stimulated the establishment of the Carnegie Commission on Educational Television and the issuance of its report, *Public Television—A Program for Action,* in January of 1967. The program proposed to:

5

1. establish a Corporation for Public Television,
2. improve facilities and provide adequate support for individual educational television (ETV) stations,
3. establish live interconnection for the system,
4. support production by local stations for "more than local use,"
5. establish program centers,
6. support local programming,
7. support research and development for improved programming and production,
8. support development for improved television technology,
9. provide means for recruiting and training personnel for public television,
10. have Congress provide overall funds through an excise tax on television sets,
11. urge the enactment of legislation to enable the Department of Health, Education, and Welfare to provide adequate station facilities, and
12. conduct studies to "develop better insight into the use of television in formal and informal education."

Public Broadcasting Act of 1967

These proposals of the Carnegie Commission are significant because they served as the basic for the Public Broadcasting Act of 1967 (PL 90-129). With the exception of the Carnegie Commission's recommendation to raise funds from an excise tax on the sale of the television sets, the other major recommendations were incorporated in the act of 1967. The act also extended the program to include noncommercial educational radio broadcasting and the Corporation for Public Broadcasting.

The introductory section of the act includes a "Congressional Declaration of Policy," which considerably expands the goals set forth in 1962. The declaration states:

1. That it is in the public interest to *encourage the growth and development* of noncommercial educational radio and television broadcasting *including the use* of such media *for instructional purposes;*
2. That expansion and development of noncommercial educational radio and television broadcasting and of *diversity of its programming* depend on freedom, imagination, and initiative on *both the local and national* levels;
3. That the encouragement and support of noncommercial educational radio and television broadcasting, while matters of importance for private and local development, are also of appropriate and *important concern to the Federal Government;*
4. That it furthers the general welfare to encourage noncommercial educational radio and television broadcast programming which will be responsive

to the interests of people *both in particular localities and throughout the United States,* and which will constitute an expression of *diversity and excellence;*

5. That it is necessary and appropriate for the Federal Government to complement, assist, and support a national policy that will most effectively make noncommercial educational radio and television service *available to all of the United States;*

6. That a private corporation should be created to *facilitate the development* of educational radio and television broadcasting and *to afford maximum protection* to such broadcasting *from extraneous interference and control.* (Italics ours.)

The act went on to delineate the purposes and activities of the Corporation for Public Broadcasting. It is authorized to:

1. Facilitate the full *development of educational broadcasting* in which *programs of high quality obtained from diverse sources,* will be made available to noncommercial educational television or radio broadcast stations, with *strict adherence to objectivity and balance* in all programs or series of programs of a controversial nature;

2. Assist in the establishment and development of one or more *systems of interconnection* to be used for the distribution of educational television or radio programs so that all noncommercial educational television or radio broadcast stations that wish to may broadcast the programs *at times chosen by the stations;*

3. Assist in the establishment and development of one or more systems of noncommercial educational television or radio broadcast stations throughout the United States;

4. Carry out its purposes and functions and engage in its activities in ways that will most effectively assure the *maximum freedom* of the noncommercial educational television or radio broadcast systems and local stations *from interference with or control of program content or other activities.*

The act authorizes more specific activities in pursuit of the purposes of the corporation. "Among others not specifically named," these are:

1. To obtain grants from and to make contracts with individuals and with private, state, and Federal agencies, organizations, and institutions;

2. To contract with or make grants to program production entities, individuals, and selected noncommercial educational broadcast stations for the production of, and otherwise to procure, educational television or radio programs for national or regional distribution to noncommercial educational broadcast stations;

3. To make payments to existing and new noncommercial educational broadcast stations to aid in financing local educational television or radio programming costs of such stations, particularly innovative approaches thereto, and other costs of operation of such stations;
4. To establish and maintain a library and archives of noncommercial educational television or radio programs and related materials and develop public awareness of and disseminate information about noncommercial educational television or radio broadcasting by various means including the publication of a journal;
5. To arrange, by grant or contract with the appropriate public or private agencies, organizations, or institutions, for interconnection facilities suitable for distribution and transmission of educational television or radio programs to noncommercial educational broadcast stations;
6. To hire or accept the voluntary services of consultants, experts, advisory boards, and panels to aid the Corporation in carrying out the purposes of this section;
7. To encourage the creation of new noncommercial educational broadcast stations in order to enhance such service on a local, state, or regional and national basis;
8. Conduct (directly or through grants or contracts) research, demonstrations or training in matters related to noncommercial educational television and radio broadcasting.

Subsequent public laws (PL 91–97 and 92–411) do not add to the goals and purposes of noncommercial educational broadcasting, but simply authorize appropriations in support of these goals and purposes.

The Law in Summary

In summary, the public laws seek:

1. development of noncommercial educational broadcasting,
2. availability to people throughout the United States,
3. participation and support by the federal government,
4. programming responsive to the needs of all people,
5. programming freedom, imagination, and initiative,
6. program production on both local *and* national levels, and
7. freedom from extraneous interference and control.

These goals are generally subscribed to both by those responsible for educational broadcasting policy and those responsible for operation of the stations and associated networks. Differences of opinion center more on questions of how the goals are to be reached and how to reconcile inherent inconsistencies rather than on the substance or intent of the goals.

Six years after the Public Broadcasting Act, the investment in educational broadcasting is approaching a quarter of a billion dollars, of which $80 million came from the EBFP. Despite this, the system is nowhere near complete, and few individual programs are funded well enough to compete with commercial production standards. The goals and funding problems of educational broadcasting are indeed formidable.

3 The Status of Educational Broadcasting

Eleven years have gone by since federal funding of educational television began; six years since the goals of educational broadcasting were set and radio became eligible for federal assistance. Today there is a working system for instructional, cultural, and informational purposes. In this chapter we will compare the status of this system with the ambitious goals it hopes to reach.

Since all of these goals are contingent upon effective contact with a broad audience, educational broadcasting can be no more successful than its distribution system. Thus, we will examine *distribution* first, comparing its extent and quality with the original target. We will then review the *programming* entering this system, considering both its nature and impact. Finally, we will look at the *funding* and administration of the system.

Throughout this discussion, it is our intent to temper the original goals with the limitations of the "real world" and to consider accomplishments in terms of reasonable expectations as well as the difficulties of comparison with a far more heavily funded industry. At the same time, we will underscore the problems that we consider most significant and areas where future assistance might prove most helpful.

Distribution

In the original Educational Television Facilities Act of 1962, the secretary of health, education, and welfare was given authority to administer grants for the construction of ETV facilities. This responsibility was then delegated to the Office of Education in 1968.[1] Thus, the initial concern was directed toward development of a radio and television distribution system. In the intervening years, as the number of stations increased, concern has turned more and more toward the salient problems of *program* goals, production, and funding.

Yet *distribution*—the array of transmission lines, antennas, transmitters, and miscellaneous hardware—is the core of electronic media. Without adequate signal strength and technical quality, without the ability to compete with the quality and pattern of commercial competition, educational broadcasting is doomed to be a voice in the wilderness, unable to meet any of its ambitious goals.

11

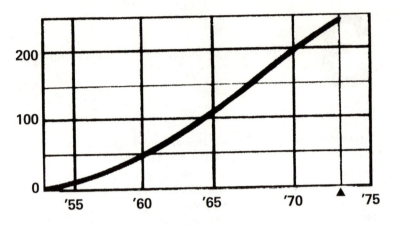

Figure 3-1. Growth in Numbers of ETV Transmitters.

Stations and Equipment

Television. The number of noncommercial educational television stations has grown steadily since 1953 (figure 3-1). As of July 1, 1973, there were 238 educational television transmitters operated by 148 licensees. Of the 238 transmitters, 94 broadcast on VHF frequencies and 144 on UHF frequencies. (Approximately two-thirds of the channel assignments currently reserved for purposes by the FCC are being utilized.)

Table 3-1 shows the distribution of these licensees across the five categories defined by the Public Broadcasting Act of 1967 (PL 90-129).

Statistics alone, however, do not provide an adequate picture of ETV facilities. Many stations are faced with problems of obsolescence, inadequate power, and marginal capabilities which are not directly reflected in numbers.

Table 3-1
Distribution of ETV Licensees

Licensee Type	Percentage of Total
Local Educational Agencies	11
Colleges and Universities	30
State Agencies	32
Nonprofit Community Corporations	27
Municipalities (owned by)[a]	
Total	100

[a]There is only one ETV station licensed to a municipality.
Source: W. Schramm and L. Nelson, *The Financing of Public Television*, Aspen Institute Program on Communications and Society (1972).

The heavy initial capital investment in transmitters, cameras, recorders, etc. cannot be considered permanent; the technology of broadcasting increases constantly and obsolescence is rapid. Even the pattern of distribution shown in table 3–1 does not present a full picture of station operation. Many stations actually are controlled by ad hoc arrangements involving combinations of agencies.

While not all problems of educational television are caused by the marginal nature of equipment and by uncertainties in funding, it is our impression that many of them are. As the manager of a small California station put it: "In many respects our most important local accomplishment is just to stay on the air." With few exceptions, there is little systematic planning of capital budgets to replace worn-out or outdated equipment; most available funds go directly for operation. This situation is usually beyond the control of individual stations. Capital improvements call for large sums which must be raised through special appropriations or major fund drives. And in many organizations that fund educational television, mere obsolescence is not interpreted as a reason for replacement.

Of the fifty station and state network operations we called, fifteen still do not have color cameras, although almost all are able to record in color from the Public Broadcasting Service (PBS). More than half of the managers felt that they did not have sufficient production facilities to maintain adequate local programming. Most had only one color-equipped studio, and in most cases there were not enough color recorders to produce local programs and simultaneously feed prerecorded programs to the air. Though some of the larger stations and state networks were well equipped by commercial standards, most color stations had just three operable cameras, and some had only two.

The lack of quality switching gear was mentioned frequently (often the switcher is a converted black-and-white unit; one station uses a converted audio switcher). Other production deficiencies were in studio space, tape-editing facilities, lighting equipment, facilities for cinematography, and graphic arts support. A few of the stations reported having elaborate mobile-unit facilities; many others would like mobile capacity, particularly where there is a single production center in a state network or where coverage of the state capital is desired. In a few cases, notably Iowa, where there is no color studio, the color mobile unit serves as the station's principal production center.

While some stations have good production equipment, there is often not enough of it to make economic sense. To set up for a single local program may require all the facilities that a station has; everything then must be rearranged for the next local program. "Every local program that we do is a 'special,' " said the manager of one station. "Setting up, tearing down, and fixing obsolete equipment is decimating our engineering budget," said a chief engineer. With the perpetual strain to get equipment, it is hardly unusual that a "famine to feast" philosophy seems to dominate some segments of educational television.

That is, when a rare station finally receives enough capital support, it tends to accumulate redundant hardware. There also is a tendency to construct buildings ("telecommunications centers") that far exceed current requirements. Such expansion is not unknown in the academic community, where the completion of a building often precedes its staffing or equipage.

Most station managers reported that their transmitters are in relatively good shape, particularly because many of them were recently constructed with funds in part made available under the Educational Broadcasting Facilities Program. A few smaller stations need high-power transmitters to cover potential viewers in semirural areas, and some older stations are in serious need of transmitter replacement. (The assistant manager of one station said that the transmitter was seventeen years old, made by a firm that no longer exists, and "We have to build our own repair parts.") Many of the stations were seeking better tower locations.

The Public Broadcasting Act called for full coverage of the population, a difficult if not impossible goal. Even commercial broadcasting, with far greater resources, does not reach *all* of the population. A number of factors contribute to the difficulty. Both television and FM depend upon "line of sight" transmission, and either rugged terrain or large steel buildings can result in "shadows" and "reflections." Persons located in a deep valley may be totally unable to receive broadcast television signals. Multiple reflections within an urban forest of steel buildings can result in a distracting array of "ghosts," and the construction of a major new skyscraper can disastrously affect the signal from a television transmitter alongside.

Almost two-thirds of educational television transmitters operate on UHF channels, where these problems are particularly serious. Furthermore, it was not until 1964 that the government required television manufacturers to include UHF; consequently, a significant portion of viewers is still unable to receive these channels. In most localities the major networks are broadcast over the earlier VHF channels, and UHF has become the "no-man's-land" occupied to a great extent by education and "independent" broadcasters.[2]

The result has not helped to change the habits of viewers, who find most of the entertainment neatly grouped together on their tuners. Despite this, ETV signals are said to be available to a nominal 72 percent of the population. Considering the UHF factors just discussed, this figure is somewhat optimistic, and real coverage is estimated at about 63 percent.

The cost of increasing the viewing population rises steadily as we attempt to cover areas of low population density and mountainous terrain. Thus, we are still far from the 100 percent goal; two states have no ETV at all except by cable systems, while in some areas several stations compete for the limited audience. Montana and Wyoming lack ETV transmitters, though both states have plans. Delaware's only television station is an educational station, though the state is saturated by nearby commercial stations. In a very few remote areas, however, such as along the Canadian border with New England, ETV provides the only U.S. broadcast television.

Radio. While direct federal support of educational radio is of more recent origin than federal support of educational television, it should not be forgotten that nearly a score of educational radio stations have long histories of service, some of them going back four decades or more. Most of these stations are operated by state universities and have been the indirect recipients of federal aid through agricultural extension services. However, thirty-seven major radio markets and substantial rural areas of the United States are not covered by educational radio, even though most of these areas are covered by educational television.

Although there are no AM channels reserved for noncommercial radio, a number of educational institutions began operation early enough to get on AM frequencies. In 1945 the FCC reserved twenty FM channels between 88 and 92 MHz for noncommercial radio, and the number of stations using these channels continues to increase rapidly (table 3-2 and figure 3-2).

As of April 1973, 550 noncommercial educational radio stations were on the air, not including the over four hundred "carrier-current" stations operated on college and university campuses. Of the 550, about 350 are low-power stations, i.e., high-school or college ten-watt stations, according to the National Association of Educational Broadcasters. The remaining two hundred constitute "large" or potentially large noncommercial radio stations. At the end of fiscal year 1973, however, only 148 of these met the minimum requirements to qualify for financial support in the form of community service grants funded from the Corporation for Public Broadcasting. These minimum requirements are: three full-time staff members, and an on-air schedule of 365 days per year, fourteen hours per day. National Public Radio estimates that 175 radio stations will qualify by the end of fiscal year 1974. A number of other stations have qualified for USOE facilities grants.

Table 3-2
Distribution of FM Station Licensees

Licensee Type	Percentage of Total
Colleges and Universities	72
Local Boards of Education and Schools	17
Churches or Religious Organizations	6
Community Nonprofit Organizations, Libraries, and Others	5
Total	100

Source: FCC Information Bulletin 21B, *Educational Radio* (January 1972).

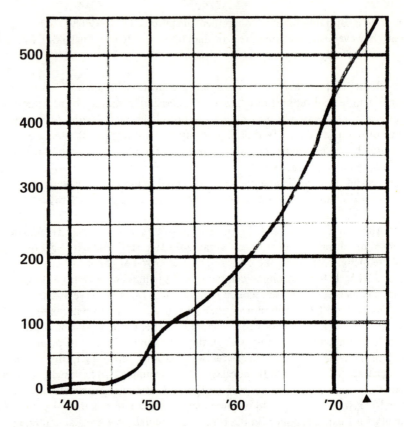

Figure 3–2. Growth of Noncommercial Educational FM Radio Stations. Source: FCC Information Bulletin 21B, *Educational Radio* (Jan. 1972).

Most operators of educational television stations feel that there is also a responsibility to supply radio services. However, in many cases radio has a low priority. Of the fifty television organizations we contacted, twenty-seven have no radio affiliates. A number of state networks, lacking adequate funds for television, have put off radio plans until budgets can be increased.

Because AM radio constitutes the bulk of listening for millions of Americans, it might be logical to consider placing an AM educational station in each of the major population centers. Only New York City, Chicago, Minneapolis, and Portland, Oregon have such a service. (The Chicago station is operated by a church seminary.)

In large parts of the United States there continue to be no radio stations that meet the community service grant criteria for financial assistance. Until recently, one could drive from Vermillion, South Dakota to Pullman, Washington, covering the entire states of Wyoming, Idaho, and Montana, without encountering such a station—more than half the distance across the continent. While community service grant requirements have not changed, the Corporation for Public Broadcasting has moved to assist a limited number of stations to meet these criteria through station expansion grants. KUFM in Missoula, Montana, received such a grant in 1974.

Educational Broadcast
Facility Interconnections

Interconnection is the lifeblood of broadcasting. A quarter million dollars per hour is a fairly common investment in commercial television documentaries, and few stations could meet this quality of programming without the economies of network distribution. A prime goal of the Public Broadcasting Act was to bring this potential to educational broadcasting, while maintaining the options for local programming and flexibility. Interconnection might also foster inter-change of local programs through development of a "party line" concept. This has not happened in commercial television, and we doubt its efficacy in ETV.

As in station and program investment, interconnection for television is far more costly than in radio. A single television broadcast requires the channel space of a thousand radio programs, and both difficulties and costs rise accord-ingly. Electronic interconnection is imperative for real-time broadcast flexibility; film or tape libraries are poor substitutes, because they are slow, cumbersome, and expensive to operate.

Most ETV stations now receive signals from the PBS interconnection system, either directly from a Bell System hookup or indirectly through a state network distribution system. In addition, many stations are subscriber members of regional systems that distribute state or regionally produced material. Despite this, only about half of the stations have adequate local equipment to take full advantage of these services.

To maintain reasonable flexibility and an ability both to produce local material and to schedule national programs for local needs, a station must have a minimum of four color videotape recorders to: record incoming material, feed recorded material to the transmitter, produce local material, and provide backup in case of equipment failure. Without this minimum hardware, the program manager is severely handicapped.

National Public Television. The Public Broadcasting Service provides the primary national interconnection facility for educational television stations. Most ETV stations now receive PBS signals in some way (figure 3-3). One hundred and ten transmitters are interconnected through long-lines facilities provided by the Bell System. A number of these transmitters serve as "hubs"

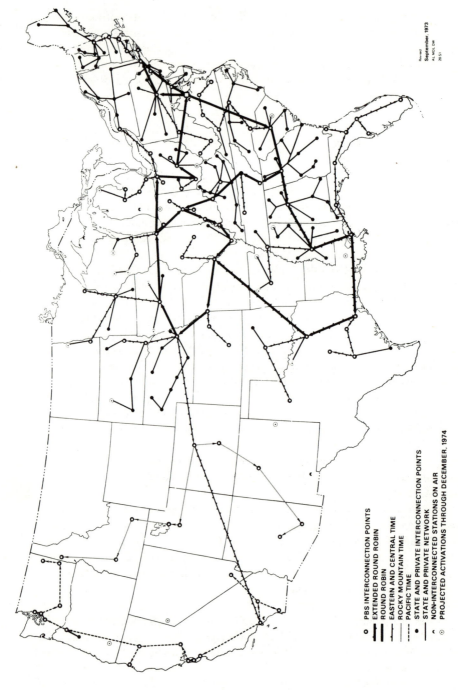

PBS INTERCONNECTION POINTS
EXTENDED ROUND ROBIN
ROUND ROBIN
EASTERN AND CENTRAL TIME
ROCKY MOUNTAIN TIME
PACIFIC TIME
STATE AND PRIVATE INTERCONNECTION POINTS
STATE AND PRIVATE NETWORK
NON-INTERCONNECTED STATIONS ON AIR
PROJECTED ACTIVATIONS THROUGH DECEMBER, 1974

Figure 3–3.   Public Broadcasting Service Interconnection System.

for state or private microwave distribution systems which reach an additional 196 transmitters. In addition, four stations provide their own interconnections to PBS. Eighteen transmitters are still not interconnected, but half of these are outside the forty-eight contiguous states. As part of the interconnection facility, PBS operates two delay centers, one in Denver, taping and retransmitting on a Mountain time schedule, and the other in Los Angeles, retransmitting on Pacific time.

National Public Radio (NPR). NPR operates a national noncommercial radio interconnection for the distribution of program material among 118 stations in thirty-six states (as of early 1972). The service is provided through "broadcast quality" 5KHz telephone lines, which are unsuitable for music or similar quality material. Material of this quality is distributed via tape recordings.

Regional Networks. Six regional networks are parts of the PBS interconnection system. All of the forty-eight contiguous states, except Delaware, Montana, and Wyoming, are included in these systems.

1. Central Educational Network (CEN) serves nine states (Illinois, Indiana, Iowa, Kansas, Michigan, Nebraska, Ohio, South Dakota, and Wisconsin).

2. Eastern Educational Television Network (EEN) serves twelve states (Connecticut, Maine, Maryland, Massachusetts, New Hampshire, New Jersey, New York, Pennsylvania, Rhode Island, Vermont, Virginia, West Virginia, and the District of Columbia).

3. Midwestern Educational Television (MET) serves two states (Minnesota and North Dakota).

4. Rocky Mountain Corporation for Public Broadcasting (RMCPB) serves five states (Arizona, Colorado, Idaho, New Mexico, and Utah).

5. Southern Educational Communications Association (SECA) serves thirteen states (Alabama, Arkansas, Florida, Georgia, Kentucky, Louisiana, Mississippi, North Carolina, Oklahoma, South Carolina, Tennessee, Texas, and Virginia).

6. Western Educational Network (WEN) serves five states (California, Nevada, Oregon, Utah, and Washington).

Eastern Educational Television Network (EEN). EEN has its administrative headquarters in Newton Upper Falls, Massachusetts, from which it serves twenty-eight licensees operating fifty transmitters. Program material is distributed over full-time video circuits leased from the Bell System and via tape exchanges. The network includes a two-way backbone route connecting Washington, Baltimore, Philadelphia, Trenton, New York City, and Boston. One-way circuits radiate from these points. Table 3–3 lists the program transmission schedule. While the facilities leased from the Bell System are available on a full-time basis, they are not used generally between 11:00 P.M. and the beginning of the next day's schedule at 9:00 A.M.

Table 3-3
EEN Transmission Schedule

Time	Program
9:00 A.M.– 4:00 P.M.	Classroom-oriented instructional material
4:00 P.M.– 6:30 P.M.	Children's programming
6:30 P.M.– 8:00 P.M.	General educational material
8:00 P.M.–10:00 P.M.	PBS evening feed
10:00 P.M.–10:30 P.M.	EEN-produced news
10:30 P.M.–11:00 P.M.	Repeat transmissions

Source: EEN.

The Eastern Educational Television Network derives 90 percent of its financial support from its twenty-eight licensees, who pay an annual membership fee. The balance represents grant money for production, which is administered by EEN. Program production is done solely by the member licensees. EEN is managed through a board of directors made up of the managers of the member stations.

Central Educational Network (CEN).   CEN, whose headquarters are in Chicago, serves nine states in which twenty-one licensees operate thirty transmitters. CEN does not lease video distribution facilities from the common carriers, but rather uses the PBS interconnect during its idle hours. The charge for this usage is minimal, consisting only of the nominal switching charge by the Bell System to isolate this part of the PBS circuit. CEN transmission consists of about eight and a half hours of scheduled transmission weekly.

CEN also operates a tape library of about 750 programs, which are distributed by "bicycling" from station to station.

The organizational structure and management of CEN is generally the same as that described for EEN.

Southern Educational Communications Association (SECA).   SECA, from its headquarters in Columbia, South Carolina, serves thirteen states in which twenty-three licensees operate about sixty transmitters. Like CEN, SECA uses the PBS interconnection to distribute material to its member licensees. SECA transmits about ten and a half hours per week.

Its organizational structure generally follows the same pattern as EEN and CEN.

Midwestern Educational Television (MET).   The MET operational center is located in St. Paul, Minnesota. MET uses a private microwave network owned, maintained, and operated by the member licensees. Currently there are six licensees operating thirteen transmitters. The system uses no leased facilities, is two-way, and has six locations from which programs can be originated. This arrangement allows recording and delay on a twenty-four hour basis.

MET incorporates a unique computer-controlled system for distribution of instructional programs. With the Computer Assisted Dial Access Video Retrieval System (CADAVRS), a teacher selects classroom television material from a catalog in advance of her scheduled use of the program. The "order" is telephoned to the system's distribution control point, where the tape is selected, given a computer "address" to indicate the requestor, and queued up for off-hours transmission. One after another, during otherwise idle hours, the programs are transmitted but ignored by the receiver until its own "address" is recognized. When the requested tape "comes up" on the queue, the address code activates the requestor's tape recorder and the material is recorded to be available in the morning for the teacher's use. When the transmission is complete, a code is transmitted which turns off the recorder.

Rocky Mountain Corporation for Public Broadcasting (RMCPB).   The Rocky Mountain Corporation for Public Broadcasting has headquarters in Albuquerque, New Mexico. It is supposed to serve seven states, but we were unable to obtain information about its operations.

Western Educational Network (WEN).   WEN serves five states through nineteen licensee members, who operate twenty-one transmitters. Program material produced by members of WEN is distributed on off hours over the PBS interconnections as in CEN and SECA. Transmission averages about one-half hour per week. WEN is organized along the general lines of the other regional networks.

Eastern Educational Radio Network (EERN).   EERN is the only formal regional educational *radio* network currently in operation. Its center of operation is Albany, New York, from which it interconnects with Boston and Springfield, Massachusetts, New York City, Philadelphia, Washington, and Richmond. EERN uses NPR facilities on off hours to facilitate recording and later replay for the member stations. EERN also bicycles tapes to the educational radio stations in Orono, Maine and Hershey, Pennsylvania.

State Networks.   At present at least twenty-eight states operate some type of local educational television distribution system (table 3-4). The organizational structure of these systems ranges from loose consortia including a variety of licensees to state agency licensees that operate all of the state's ETV transmitters.

Other states operate partial systems, i.e., distribute to one or more but not all of the transmitters in the state. Minnesota currently distributes program material to three of its four transmitters through the Midwestern Educational Television Network. Other informal arrangements exist; often they are put together to facilitate coverage of a major event, such as a special session of the state legislature.

Table 3–4
State ETV Networks

State	Type of Licensee	Number of Licensees	Number of Transmitters	Number of Translators
Alabama	State Authority	1	9	3
Connecticut	Community	1	3	0
Georgia	State Board of Education	3	10	4
Hawaii	University	1	2	5
Indiana	Composite	2	2	0
Iowa	State Authority	1	3	3
Kentucky	State Authority	1	13	6
Maine	University	2	4	4
Maryland	State Authority	1	2	1
Minnesota	Composite	3	5	6
Mississippi	State Authority	1	6	0
Nebraska	State Authority	1	7	4
New Hampshire	University	1	5	1
New Jersey	State Authority	1	4	0
New York	Composite	10	11	4
North Carolina	University	1	7	0
Ohio	Composite	8	9	0
Oklahoma	State Authority	1	2	0
Oregon	Composite	2	2	1
Pennsylvania	Composite	7	8	10
Puerto Rico	Board of Education	2	2	1
Rhode Island	Board of Education	1	1	0
South Carolina	State Authority	1	5	1
South Dakota	State Authority	1	3	10
Tennessee	State Board of Education	2	4	0
Utah	Composite	4	4	24
Vermont	University	1	4	3
West Virginia	State Authority	1	2	0
Totals		62	139	91

Sources: Public Broadcasting Service, L.T. Frymire, *Study of State Public TV System* (1969).

Means of Extending Broadcast Coverage

Translators. A "translator" is a combination television receiver and low-power transmitter designed to rebroadcast the signal on a new frequency. Thus, this "slave" transmitter can reach areas not reached by the station's main transmitter.

As of June 1973, forty-eight stations owned and operated 138 translators directed at areas unable to receive the primary broadcast signal. In addition, twelve construction permits had been granted by the FCC. These translators, for the most part, are 10 watts in power, but there is a scattering of 100 and 1000 watt units.

Although many of the stations we surveyed employ translators to boost their signals, the value of translators, as understood by the individual station managers, is by no means clear. Some said they add significantly to potential viewing audiences; others said they do not, because of tuning problems and low picture quality. Still others said they "help fill in the vacant spaces." (A notable translator success story is that of Utah, where twenty-four translators bring the Salt Lake City channel to nearly the entire state.)

Cable Television (CATV) Systems. Cable communication is an alternative means of distributing broadcast signals and extending their range. As such, it can and does extend the noncommercial educational TV signal just as it does the commercial counterpart. Thus, a portion of the 72 percent of the American population that has an educational broadcast signal available receives that signal via cable. There are, however, no definite data regarding the use of CATV as a means to extend ETV coverage. As might be expected, we found in our telephone and site interviews that use of CATV to extend the signal varied primarily with the terrain and the size of the community served. Some ETV managers did not know how many, if any, CATV systems carried their signals, and others worked closely with over one hundred systems to insure such rebroadcast. In Los Angeles, for example, because of the terrain and the fact that all the radio broadcast frequencies are assigned, as are most of the TV frequencies, cable is a sought-after means of distribution of the ETV signal.

Station experiences with cable television vary widely. There is broad distribution in states, such as Oregon and Pennsylvania, with well-developed cable systems. In a few areas, cable provides the only access to public television. (For example, the only service to residents of Montana comes via cable from Spokane and Salt Lake City.) Significant audiences in Canada and Mexico receive public television via cable from stations in bordering U.S. states. There is frequent engineering contact between cable interests and station operators who furnish information on how best to pick up the local station signals. However, among the stations we surveyed, only one or two had furnished videotapes for replay at a cable system's headend.

A few of the operators of black-and-white stations said that cable interests preferred going out of their local communities to get public television signals that could be relayed in color. The FCC requires that all new cable operators carry nearby educational channels. However, the problem of persuading older cable companies to add a local educational channel where cable capacity is being fully used is a vexing one, particularly to the infant Maryland state network. While all of the station managers we interviewed were aware of the potentialities of cable, only about half reported anything more than informal contacts with cable companies.

A few cable systems are already *originating* educational material for the schools in their franchise area and for their subscribers. In these systems, which provide the educational channel required by the FCC in newer systems, programming is being developed and used to supplement existing educational resource material.

The National Cable Television Association has an active Educational Cable Television Committee whose charge is to explore the use of cable for dissemination of educational material.[3] This group has identified successful programs in sixty-six communities in thirty-two states distributed over CATV systems. In addition, sixty-five colleges and universities use CATV for distribution of educational material or for training. Examples of such efforts are those in:

1.	Dougherty County, Georgia	Instructional programming and local public affairs.
2.	Casper, Wyoming	Film programs distributed to school on cable channel, sports and review courses.
3.	Bainbridge, New York	School-owned studio which produces elementary school.
4.	Abilene, Texas	General educational programming in cable operator's studio.
5.	Hagerstown, Maryland	Interconnection between school's closed-circuit TV and cable system.
6.	Moab, Utah	Studio interconnected to cable headend with capability for live, film, and tape formats.
7.	Ellensburg, Washington	Central Washington State College TV system to cable system.

While these examples represent only a few of the three thousand or so cable systems in operation, they do indicate an initial interest in the use of cable to distribute educational material.

Instructional Television Fixed Service (ITFS). In 1963, the Federal Communications Commission established a program to provide licenses for *multiple* video channels for ITV purposes in the 2500–2690 MHz frequency range. The purpose was to relieve conventional broadcast channels of instructional traffic where materials are not intended for home use. A single ITFS system can simultaneously transmit on four separate audio and television channels.

An ITFS system operates on low power output and has a maximum useful service range of about twenty miles. The cost of construction and operation is less than the cost of the transmitting equipment for standard broadcast television, although an expensive special receiving antenna and conversion device are required so that a conventional TV receiver can be used. By August 1973, there were 152 ITFS systems in operation, providing programming on about 450 channels. In addition, thirty-three systems have been awarded construction permits but were not yet on the air.

The purpose and functions allowed to ITFS systems are:

1. To provide a means of transmitting instructional and cultural material in visual form with associated aural channels for the purpose of providing formal educational and cultural material to students enrolled in accredited public and private schools, colleges, and universities.
2. To provide in-service training and instruction in special skills, safety programs, and professional continuing education.
3. When not in use for disseminating educational or cultural material, the system can be used for transmission of information pertaining to school administration. (Stations *are not* licensed whose *primary* purpose is the transmission of administrative traffic.)

In addition to its function of distributing specialized materials to limited audiences, ITFS provides a means to transmit instructional material to classrooms without the level of expense required for a conventional educational TV station. When used to relay material from the local ETV station, ITFS provides greater program flexibility and greater freedom in scheduling the use of supplementary program material. However, transmission economies touch only one aspect of the instructional television system; the special costs of reception and program production remain. Hence, there has been only limited interest in and modest growth of ITFS.

Programming

Lord Reith, the father of British broadcasting, repeatedly stressed the strategy of using radio and television as a cultural catalyst, constantly raising the abilities of audiences to discriminate in favor of things worthwhile. This, he said, should be accomplished in Britain "by brute force of monopoly" through complete control by a public corporation. While the success of such a policy may be debated, there was no such original strategy in the United States, where the public media must compete with other uses of time to unlock the door to the cultural alternatives they are convinced they can bring.

Table 3-5

Television Service Costs Distributed on a Per Capita Basis—By Country

Country	Service	Expenditure Per Person
Canada	CBC–All CBC TV including commercial programming (1970/71)	$ 7.70
United Kingdom	BBC–Totally noncommercial (1970/71)	3.29
Japan	NHK–Television share of NHK operation (1971/72)	2.90
United States	Commercial TV–Station and network revenues (1970)	13.71
	Network revenues only (1971)	7.32
United States	Educational Television–Station revenues (FY 1971)	0.69

Source: W. Schramm and L. Nelson, *The Financing of Public Television,* Aspen Institute Program on Communications and Society, Palo Alto, California (1972).

The challenge is not to win a rating war. Rather, it has several parts: to supply a worthwhile service, to attract an audience, and to identify and satisfy the needs of people in ways that are not available elsewhere. As public broadcasting approaches national availability, issues of programming are becoming increasingly important. They are recognized to a greater or lesser degree by all of the stations we have contacted.

Production Costs

The per capita U.S. expenditure for public television (table 3-5) is far below that of other countries with well-established public services and insignificant alongside the expenditures for commercial TV, with which it must compete. Despite the fact that direct operating costs for public TV amount to more than 75 percent of total expenditures, most programs are produced on budgets that average about 25 percent of their commercial counterparts.[4] While there is little definitive information, Schramm and Nelson estimate that program costs range from $2500 to $5000 per hour for locally originated programs to $100,000 for special events.[5] A comparable commercial program could easily represent five times this investment in production and promotion. Thus, we ask public television to provide innovative education and viable alternatives to commercial broadcasting with minimal funding.

Characteristics of Educational
Television Schedule

In April of 1972, the Corporation for Public Broadcasting undertook to summarize the volume and type of programming produced by and for ETV by surveying the activity in a representative week.[6] At the time there were

Table 3-6
Distribution of Locally Produced Program Material by Licensee

	Percent Produced Locally
University	12.8
State Authority	16.6
School Board	24.5
Community	15.5

220 stations on the air. During the week, an average of seventy-one hours of programming per station was transmitted to educational TV audiences. Comparing this to 1970 data, the report indicates a 16 percent increase in the number of stations, and about 28 percent more total hours of broadcasting. Most data in the survey are based on transmissions by broadcasters rather than by stations, since a single broadcaster (such as a state network) might transmit the same program on several stations. A total of 149 broadcasters were counted.

The transmission schedule averaged 68.5 hours for the week per broadcaster, about 11.7 hours on weekdays and 4.8 hours on the weekend. About 41 percent of the broadcasters transmitted all seven days of the week, 45.6 percent transmitted six days, and 13.4 percent only five days. Of 10,202 total hours broadcast, 31.1 percent were produced and broadcast in black and white, 63 percent were broadcast in color, and 5.9 percent were produced in color and broadcast in black and white. Of that same number of hours, 3.2 percent were broadcast live, 33.6 percent were served by an interconnection, and 63.2 percent were transmitted from recordings.

Sources of Programs

Although by far the largest share of material shown by U.S. educational television broadcasters is provided by PBS (figure 3-4), the percentage of local program production time varies greatly (table 3-6). During a typical 1972 week, the three broadcasters with the highest budgets each produced about 20 percent of their entire program schedule time locally, while the two lowest budget broadcasters each produced only 9 percent of such material locally. Our interviews indicate a range from 35 percent to *no* local programming. It would be a mistake, however, to ascribe this difference solely to inadequate equipment. In our opinion it reflects the budgets, logistics, talent, and sophistication needed to produce programs in significant quantity and quality. Even commercial stations are hard pressed to compete with network standards, and the desire for universal local production must be tempered with reality.

In the CPB survey cited above, the broadcast hours were directed predominantly toward the adult age group (45.5 percent), while 25.3 percent of the hours were oriented toward the six-to-twelve age group. Twenty percent

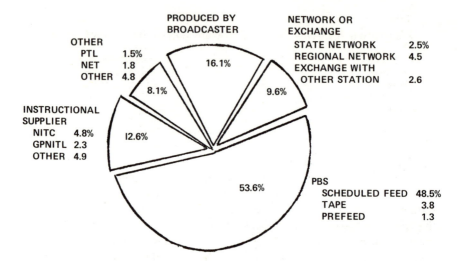

Figure 3–4. General Programming by Source. Source: *One Week of Public Television, April 1972*, p. 16.

were for the under six group, 7.9 percent for the twelve-to-eighteen group, and 1.4 percent specifically for the young adult. The CPB survey also made a breakdown (figure 3–5) of broadcast hours according to their division into four general program types.

A comparison of PBS programming with commercial network distribution (table 3–7) emphasizes the cultural aspect of public TV.

This comparison reflects a significant bias that cannot go unmentioned. Public broadcasting has been accused of being programming for a white, upper-class, intellectual minority. Despite programs such as "Black Journal" and "Sesame Street," a strong case can be made for this point of view. What is "cultural" for one minority group may be ignored by another minority as irrelevant. Table 3–7, for example, shows no "cultural" or "children's" programs on commercial TV. (Apparently the Saturday children's shows are classified elsewhere, along with "Lassie" and "Walt Disney.") We suspect that a film like *Alexander Nevsky* might be considered an "antique" feature film in the top column, but a "classic" cultural program in the bottom one.

Percentage figures tend to minimize the almost 10:1 ratio of commercial to educational broadcasting. The report notes, for example, that "less than 4% of all commercial network programs were news, public affairs, or documentary programs in prime time." This implies that more time is needed—yet the table ratios represent an average public affairs time for each network of 10 hours versus 8.7 hours on public broadcasting.

Still more striking is the inclusion of daytime serials, etc. in the com-

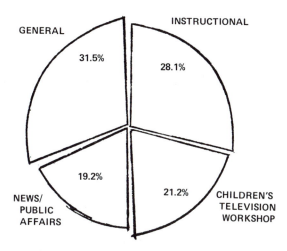

Figure 3-5. Percentage of Total Hours (10,202) Broadcast, by Type. Source: *One Week of Public Television, April 1972*, p. 22.

Table 3-7
Comparison of Commercial Program Distribution with PBS
Program Distribution

Three Commercial Networks	
253 hours distributed	
Type	*Percent of Total*
Drama/Adventure	11.8
Comedy/Variety	18.2
Feature Films	10.6
Daytime Serials	15.8
Quiz/Audience Participation	13.8
Sports	9.3
News/Public Affairs	12.4
	100.0
Public Broadcasting Service	
29 hours distributed	
Type	*Percent of Total*
Cultural	9.5
Public Affairs	30.1
Performance, Nonmusical	13.8
Performance, Musical	6.9
Children's	39.7
	100.0

Source: *One Week of Public Television, April 1972*, p. 24.

mercial column, but exclusion of daytime instructional television from educational broadcasting! Most *housewives* would surely consider instructional television a daytime cultural wasteland! Figures do not lie, but they certainly can mislead.

The Two Types of Services

For the purposes of examining the activity in educational broadcasting, we have arbitrarily defined *instructional TV* (ITV) as that programming broadcast roughly between the hours of 8:00 A.M. and 3–4:00 P.M. which is oriented toward classroom use. We have defined *public TV* (PTV) as that oriented toward home reception—i.e., the broadcast schedule transmitted after about 4:00 P.M. and until signoff (usually between 11:30 P.M. and 12:30 A.M.)

It is difficult to devise more clear-cut definitions for ITV and PTV, and we recognize the overlap of categories. For example, college credit courses are clearly instructional material: oriented, in many respects, toward classroom-like utilization; but to be received on a TV set in the home. "Sesame Street" and "The Electric Company," while oriented toward the home viewer, are used extensively in the classroom. Some stations repeat these programs several times as staples of both their classroom and home services. This practice can result in a surprising degree of local penetration, such as in Chicago, where "Sesame Street" is reported to reach 95 percent of all black households. College-level ITV material appears frequently in the late afternoon or evening hours, as does the widely used adult high-school series, "Your Future is Now." A few stations schedule in-service ITV for evening hours, when teachers can view the material at their leisure.

Instructional Television.   Despite the fact that approximately half of the financing for public broadcasting derives from state funds, there is a dearth of recent studies or analyses of the utilization of instructional television. In many states, statistics on pupil utilization are obtained from time to time, but there is little uniformity in the way the states keep their records, and state-to-state comparisons are impossible. Sometimes this record keeping is viewed as a function of the state department of education, and station managements are not even familiar with the results.

In the few cases where we were able to obtain useful data, "The Electric Company" was the leader in classroom utilization, followed closely by "Sesame Street." Primary-level materials that might be termed "general learning enrichment" were popular, as were elementary-level science, speech improvement, literature, and handwriting.

A notable exception to the dearth of survey material is a 1972 study conducted for the North Carolina state network. Quoting from other network documents, this study makes a series of comments on the metamorphosis of ITV since it originated in North Carolina in 1957. From our observations, we believe these comments also apply very well to current trends in other states:

> The fourteen years since the beginning of broadcast instruction have brought enormous change in both theory and practice. The original goal of providing *total instruction* via TV has changed to a *supporting role:* providing children an enriched environment for learning. A second early goal of *quantity:* teaching hundreds of students in a class, has changed to *stimulus* where 25 to 35 students view the programs together and each student sees the program and responds to it as an individual. Still a third original goal of using space not ordinarily used for instruction has been replaced by the concept of carefully integrating television into the student's normal learning environment. Here it is possible to prepare for the television programs, to view them, and to follow up the programs without interruptions. Most teachers prefer to use ITV *in* their classrooms.
>
> In curriculum too, instructional television has undergone a major transformation. The intent is no longer to provide total teaching of any subject. Televised lessons are no longer designed sequentially—that is, the material covered in any one program does not require the student to have viewed all or even some of the preceding programs. Instructional series are constructed only to provide the classroom teacher with an aid to the epistemological process; a tool to be used with discrimination and not in any way a substitute for classroom instruction.[7]

Interviews of fifty television broadcast managers gave us an overall view of instructional programming on 138 public channels (including network outlets).

All but one of the fifty station managers reported participation in instructional television. (The exception was a station operated with community funds which presently has no formal contact with local schools.) Although a number of the station managers did not believe ITV in practice was living up to its potentialities, almost all have three or more hours of daytime series aimed at the kindergarten through primary grades. There is somewhat less programming of formal junior-high-school material, and only about half of the stations have more than one or two series for high-school students. The reason most frequently cited for lack of participation in high-school classes was scheduling difficulties. (To reach the high-school group, a series usually must be shown at a certain hour, and this time usually varies from school to school.) In a few cases where the same management operates two ITV channels, the second channel carries secondary-school programs repeated frequently. A few stations (notably the Vermont state network) take special pains to show secondary-level programs that are not formalized but are keyed to timely local and national events. Similarly, a few stations have special news programs for elementary-school viewing.

Twenty-six of the fifty station managers reported no formalized courses for college credit, although many were interested in the "open university" concept, and three had prepared formal open university proposals. Of the twenty-four offering college courses, only five have extensive programs that could contribute measurably to a student's academic standing. Eleven stations present three or more courses, mostly for adult extension education. Eight present one or two college courses a year, often as tokens of the stations' academic relationships. The lack of college-level participation often is due to disagreements over accreditation procedures and, as one station manager declared, "the general snail's pace of the academic community."

Provision of special ITV for professionals and for trade and business groups depends on the financial support of a station and on its degree of penetration within its community. These services are most popular where there are two available channels or where, in a few cases, the stations cooperate with existing ITFS systems.

There was considerable disagreement among station managers regarding the value of locally produced ITV as opposed to the value of material obtained from National Instructional Television (NIT), the Great Plains National Instructional Television Library, or other libraries and commercial services. Some prefer to produce ITV locally, while others prefer the national services for materials that are not intrinsically parochial, such as series on state history and government. Even in the same geographical area there are pronounced differences. While South Carolina, Alabama, Mississippi, and Georgia all are extensively involved in local production, Kentucky produces almost no local classroom ITV. On the average, each of the fifty television organizations we contacted produces 28 percent of its ITV material.

This lack of uniformity reflects both the still exploratory nature of instructional television and a lack of integration within the general framework of classroom teaching in the United States. This is not entirely bad; since ITV has not become institutionalized there still is room for experimentation. In particular, "Sesame Street" and "The Electric Company" have pioneered an area where ITV is not formalized course material and resembles public affairs programming in that it is keyed to changing events. "If I can get to the consumer when he's about to buy a car, or the kid when his friends are trying to get him to use amphetamines—then I will have filled my ITV function best," one station manager told us.

However tenable, this viewpoint still reflects the fact that ITV is not yet a part of routine practice in the school systems. An approach is needed that both encourages innovation and provides formal material that is of uniform high quality.

The unifying patterns of National Instructional Television and the Great Plains National Instructional Television Library may provide such an approach in the future. NIT, located in Bloomington, Indiana, and Great Plains, located in Lincoln, Nebraska, have similar library functions but different production philosophies. NIT prefers to form consortia of state school systems which plan and then finance the production of ITV series through grants. The Association for Instructional Television (AIT) has been formed to handle the production functions of NIT. Great Plains obtains ITV series from the producers of the original materials (usually local stations or state networks), sometimes rewriting the printed teachers' guides for national distribution. Other organizations, such as the Eastern Educational Television Network (EEN) and Telstar (a spinoff from the Minneapolis-St. Paul educational television station, which produces and distributes adult-level material), are having a similar unifying effect.

NIT is deeply committed to basic planning and content control by the educational community that will use its material. A managerial group representing the member school systems establishes a list of high-priority ITV materials and these form the basis of a series of prospecti. Each prospectus is then submitted to the members of the group, who then may choose to participate by contributing to the cost on the basis of TV-student population. The participants in a particular series also send representatives to an executive committee that supervises the detailed content of the series. In this way NIT is pioneering an unusually responsive system to produce ITV material.

During *One Week of Public Television,* 2869 hours of instructional programming were transmitted in the United States distributed into the categories shown in table 3–8.

The 2869 hours broadcast represented 8259 instructional programs, or an average of about 17.5 minutes per program. The 2869 hours were distributed by grade level as shown in table 3–9.

Sources of instructional programming are shown in table 3–10.

Table 3–8
Percentage Distribution of ITV Program Material by Type

Subject Area	Percent of Total Hours Broadcast
Social and Behavorial Sciences	26.5
Language Arts and Literature	16.4
Physical Sciences	14.9
Music and Dance	7.2
Fine Arts	5.4
Mathematics	5.2
Health, Safety, and Physical Education	4.3
Foreign Language	3.8
Education	5.1
All Others (less than 2% each)	11.2
	100.0

Source: *One Week of Public Television, April 1972,* p. 29.

Table 3-9
Percentage Distribution of ITV Program Material by Grade Level

Grade Level	Percent of Total Hours Broadcast
K–3	23.6
4–6	33.5
7–9	14.2
10–12	13.6
College	5.0
Adult	9.6
No Response	0.5
	100.0

Source: *One Week of Public Television, April 1972*, p. 29.

The portion of hours of ITV programming produced locally decreased evenly from 1964 to 1970 from a level of about 56 percent to about 27 percent. The portion rose to about 32 percent in 1972, but some of this rise may be due to the introduction of a different way of determining units of measure.

Of the 8259 instructional programs broadcast, 84.3 percent were recorded, 14.9 percent were interconnected, 0.9 percent were live. Of the total, 76 percent were broadcast in black and white, 47.5 percent were repeated within seven days, and 60 percent had been shown in prior years. About 72 percent of the instructional programs were supplemented with materials for the teachers; 8 percent were supplemented with material for the student. Two percent were designated by the broadcaster for adult continuing education.

Public Television. A keystone of federal policy for public television, frequently expressed by the Office of Telecommunications Policy, Executive Office of the President, is the need for stations to respond to local needs rather

Table 3-10
Sources of ITV Program Material

Source	Percent of Total Hours Broadcast
Local Production	31.9
Direct Exchange	5.8
State Networks	6.3
Regional Networks	7.1
National Sources (PBS, NET, PTL)	5.2
National Instructional Television Center	16.4
Great Plains National Instructional Library	10.2
Other Instructional Distributors	14.2
Other	2.9
	100.0

Source: *One Week of Public Television, April 1972*, p. 32.

than function merely as outlets for a flow of material originating in national production centers. There is no question that this already is the case. The constraints under which local stations operate vary so widely that diversity rather than universality is the predictable pattern. Rather, national programming provides services, skills, and scale of production that could not possibly be provided at a local level. One station manager echoed the opinions of others when he said:

> I like public television as a local service because it gives me an opportunity for creativity which no commercial station in a local community can provide. . . . However, though we emphasize the local view, it is the *national system* that gives us high-level programming which will attract the viewers, which we could not possibly produce ourselves.

A few local stations with limited resources rely almost exclusively on national public television services. As far as we can determine, it would be impossible for these stations to operate without the national programming. For others, national programming serves as the stations' biggest drawing card for both audience and local fund raising.

Television stations in the UHF channels face an especially difficult problem in attracting viewers. This difficulty was mentioned as crucial by the operators of several state networks. The problem often is compounded by a reluctance of CATV operators to assign channels to the newer UHF stations in areas where there are already many commercial channels (in spite of FCC requirements to carry at least one ETV station). One possible solution is to develop a program schedule that viewers will look for, placing heavy emphasis on news, sports, and local public affairs. The reporting of news events is not a function of most educational television stations, though educational radio stations often place heavy emphasis on their news departments. Of the fifty television organizations (individual stations and state networks), only nine have active news staffs; however, all nine regard news as an important element of viewer service. Similarly, only a few stations make special efforts to cover spectator sports; yet, where there are sports departments, they are highly regarded. All of the stations stress local public affairs, with coverage ranging from discussions of local issues, to open-line viewer participation, to gavel-to-gavel presentation of meetings of school boards, city councils, and state legislatures. All would like to do more local public affairs.

There is a rich variety of specialized programming to satisfy local needs: public-access programming through provision of an open channel; health, physical education, and homemaking information; presentations of fishing, hunting, gardening, and boating information; magazine programs featuring film on local events; a wide range of documentaries and minidocumentaries; and special programs for minority groups. While the variety is almost endless and reflects the ingenuity of station staffs, it was apparent that these programs frequently were inspired not by the stations but by outside groups who came to the stations asking for a service, often with financial support in hand. In particular, among the older established stations there seems to be a lack of interest in providing services, however badly needed, unless someone else has the money to get the job done. There also seems to be a tendency to relegate programming aimed at minority groups to certain schedule segments (such as the half-hour weekly "For Blacks Only") rather than to serve minority groups across the board. One station manager provided the rule-proving exception when he said, "We purposely have no black program; our black producer-director wants to be effective across our entire program schedule."

The Public Television Audience.   There have been few local efforts to determine the size of the public television audience. Station managers sometimes get a rough idea of their ratings by taking a peek at the results of American Research Bureau (ARB) surveys paid for by their commercial counterparts. Only when there are series such as "Sesame Street," "The Electric Company" or "ZOOM," or well-advertised specials, are the figures uniformly large, though the station managers we contacted reported high response to the Watergate hearings. In some areas, local news, sports, and public affairs programs command sizable audiences, as do some PBS cultural programs.

"Because our viewers are *motivated* viewers who watch us for a particular purpose, we are not interested in numbers" was the most frequent comment by station managers we interviewed. This is not entirely an excuse for lame programming. Like the BBC's celebrated "Third Programme" or Britain's second public television channel, public broadcasting in the United States is basically motivated toward quality, not quantity. However, it is our impression that few station managers would not be thrilled to achieve an evening ARB rating of 13—the level attained by the National Geographic specials on commercial television.

The Corporation for Public Broadcasting recently funded a study of national patterns of viewing educational television.[8] This study included two national samples, comprising 149 sampling points in areas where educational television can be received. The study team divided the country into four areas and attempted to assess availability as well as viewing patterns. More than two thousand persons over fifteen years of age were interviewed. Most of the following information is derived from this study.

Who *can* watch educational television? There are approximately 65 million television households in the United States, about 96 percent of *all* households (figure 3-6). Educational television is available to about 72 percent of the TV households: a total of 47 million households containing almost 150 million people (figure 3-7).

Of the four geographical areas, the East and West have the highest ETV saturation, although there is considerable variation, particularly within the West. Eighty-five percent of the TV households in the East and West can get

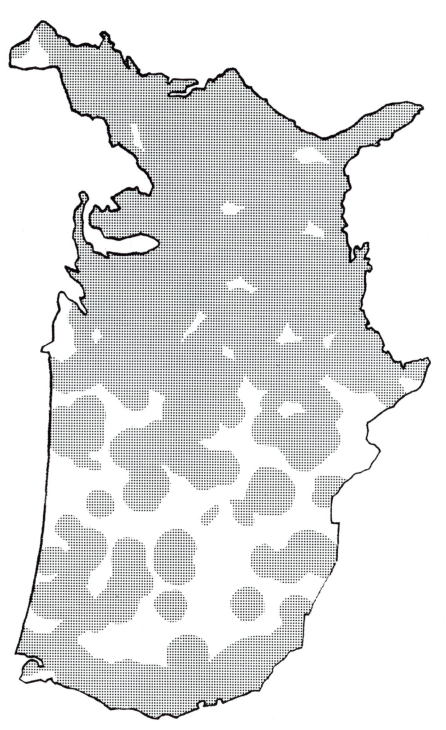

Figure 3–6. Commercial Television Coverage.

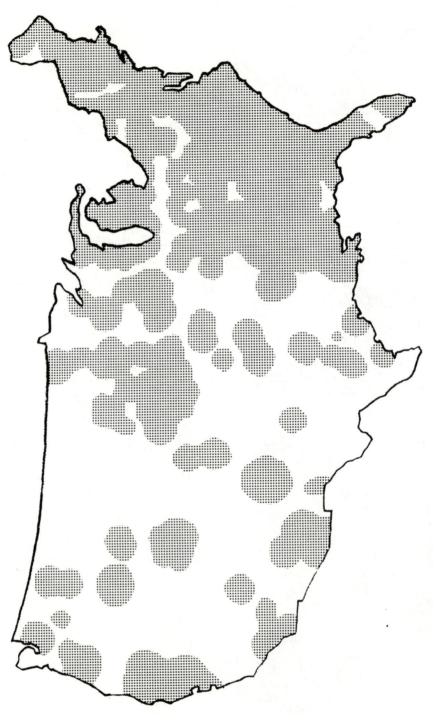

Figure 3–7. Noncommercial Television Coverage.

educational television. In the Midwest and South, ETV is available to only about 60 percent of the television homes. As would be expected, the study indicates higher availability in large metropolitan areas.

The actual availability of educational television is a function of: the number of stations, the signal strength transmitted, and the channel location. There were 695 commercial TV transmitters on the air at the beginning of 1973. There were 235 noncommercial educational TV stations on the air at the same time. Of the educational TV transmitters, roughly 60 percent broadcast on UHF channels, compared to only 28 percent of the commercial TV transmitters.

While most (about 60 percent) of the ETV transmitters broadcast on UHF, about 57 percent of the educational television viewing households receive ETV on VHF. Only 72 percent of these households in UHF areas have sets *equipped* to receive such transmission. With these limitations in mind, the Harris study concludes that only about 63 percent of all TV households *can actually receive* educational television programming. Thus, the probable potential coverage of ETV is about 41 million households, or about 130 million people.

Who *does* watch educational television? The Harris study indicates that during a one-week survey in the area where ETV is available, approximately one-third of the population actually watched educational television at some time for an average of approximately two hours (tables 3-11 and 3-12).

Table 3-11
Hours of Educational Television Watched Per Week

	Median Hours/Week	
	1971	1970
Total	1.9	1.5
Cities	2.2	1.5
Suburbs	1.9	1.5
Towns	2.0	1.8
Rural	1.2	2.0
16 to 20 year olds	1.6	0.8
21 to 29	2.2	2.0
30 to 49	1.7	1.5
50 and over	2.0	1.8
Men	1.7	1.4
Women	2.1	1.7
White	1.8	1.6
Black	2.1	2.2
Less than high school	1.8	1.2
High-school graduate	1.8	1.6
College graduate	2.1	1.8

Source: *The Viewing of Public Television-1971*, prepared for The Corporation for Public Broadcasting, Louis Harris and Associates, Inc., Study No. 2135 (November 1971).

Table 3-12
Educational Television Programs Watched by Interviewees

	Total Percentage Watching
By Specific Program (1971)	
Sesame Street	27
Misteroger's Neighborhood	7
Firing Line	5
Boston Pops	4
French Chef	3
Soul	3
Masterpiece Theatre	3
First Churchills	3
David Susskind	3
American Dream Machine	2
Black Journal	2
Army Basic Training	2
Civilization	2
Hodge Podge Lodge	2
Advocates	1
Forsythe Saga	1
By Type of Program (1971)	
Education	8
Discussion/Interviews	7
Musical Programs	7
Plays	5
News	4
Movies	4
Instructional	3
Documentaries	3
Sports	20
Other	20
Don't Know	13

	Percentage By Year		
	1971	1970	1969
Watched in last week	34	26	21
Watched in last six months	51	43	29

Source: *The Viewing of Public Television-1971,* Prepared for The Corporation for Public Broadcasting, Louis Harris and Associates, Inc., Study No. 2135 (November 1971).

It is interesting to note that the median hours increased from 1.5 hours viewing per week in 1970 to 1.9 hours per week in 1971. This amounts to about 12 percent of all weekly TV viewing by those surveyed in 1971. The median hours of all TV viewing (commercial and noncommercial) for the week was 16.3 hours. Thus, the Harris study concludes that educational television is used selectively and represents a relatively small percentage of total TV viewing time.

The study also concluded that respondees felt:

1. Commercial television did not appeal mainly to those with higher education.
2. Commercial television appealed to a larger segment of the community.
3. Both commercial and ETV contained worthwhile programming for minorities.
4. Both were willing to take risks with unusual programming.
5. Commercial television more often dealt with controversial subjects.
6. Commercial television was more often run by competent professionals.
7. Commercial television more often presented a biased point of view.

Innovations in TV Programming

While each of the stations we contacted has some feature that makes it unique and may well be worth introducing elsewhere, there were nine facilities to which, on the basis of our interviews, we would assign high "venture" ratings. That is, these stations seemed most willing to take innovative steps to serve their publics. The nine are:

WTHS, Channel 2, and WSEC, Channel 17, Miami, Florida, for innovative instructional television programming to serve a rapidly developing community. Operated by Dade County Schools, both stations provide educational services, WTHS sharing its channel with community station WPBT. With small staffs and limited facilities, the stations carry extensive instructional television programming as well as school-related community activities, special programs for Cuban-Americans, model cities programs, and programs for adult education. Channel 2 serves elementary grades, and Channel 17 serves junior and senior high school, with adult education in the evenings. Particularly for secondary schools, courses are repeated frequently to achieve saturation. In addition, there are two ITFS channels that serve all schools. In the fall of 1973 a two-way interconnected ITFS system was scheduled to be inaugurated, with connections to several universities in its 120-mile service area. The station management also is experimenting with simultaneous instructional television programming in Spanish and English, with the second language carried on an FM radio subcarrier, and with the concept of providing personalized audio instruction from a central location with ninety-six audio channels in the spectrum of a single video channel.

WDCN, Channel 2, Nashville, Tennessee, for innovative approaches to upgrading its service. Operated by the Nashville Metropolitan Board of Education, this station was in the position of a number of educational television stations whose facilities were acquired several years ago. The original black-and-white equipment, most of it leased from the state, was out of date and "going downhill fast." Through a combination of funding approaches, the station management is seeking to update its equipment and build a 1.7 million dollar educational telecommunications facility. These approaches include a channel swap with a commercial station (Channel 8 for Channel 2), a school bond issue, and application for an HEW EBFP grant. Aiding the station is the Nashville Public Television Council, which is a promotional and policy organization and helps coordinate the fund-raising activities. The council's second auction in 1973 raised $71,000, a 40 percent increase from its first auction. Initial improvements will be the erection of a new transmitter. This will be followed by acquisition of color studio equipment and construction of the telecommunication center.

Maine and New Hampshire public television networks (WMEB, Channel 12, Orono, and WENH, Channel 11, Durham), for innovative approaches to local public affairs programming. Both of these networks receive somewhat more than half of their operating funds from state government sources. Both have highly successful nightly news and public affairs programs which focus on state activities. The New Hampshire network presents a weekly program on citizens' activities, shows weekly highlights from state educational institutions, and does extensive coverage of legislative activities, including a recent full day at the state senate. Four to six film documentaries, often on public issues, are constantly in production on a contract basis. In spite of the fact that it does not yet have local color facilities, the Maine network has become the major outlet for news from the state capital when the legislature is in session, using a "jury-rigged" microwave unit. The network runs public access "By the People" programming following the WGBH format and was increasing this coverage in the fall of 1973. A series about work and career opportunities in Maine is highly successful, as is a weekly magazine program for the elderly (in 1973 running out of funds). Both state networks contribute programs to the Eastern Educational Television Network and do a limited amount of production for PBS. Both are working on open university proposals. The Maine network is proposing a weekly Canadian-American public affairs program, with consideration of common problems and viewer participation from both countries. The New Hampshire network is negotiating with Ontario to set up a weekly French-English bilingual educational program.

KAID, Channel 4, Boise, Idaho, for innovative local public service and political coverage. Only one and a half years old, KAID provides an example of what can be done by creative management on a $210,000 operating budget (1972-73). The station started extensive local programming from a commercial studio before its own studio was furnished. During the last state legislature, the station programmed a daily fifteen-minute and weekly half-hour review. There was gavel-to-gavel coverage of two major public hearings and coverage of two of the governor's messages. (Much of this programming was carried on the Pocatello educational television station and on Idaho commercial radio stations.) Extensive election coverage featured an offer of free air time to all state legislature candidates in the station's coverage area (34 out of 75 accepted). There is a weekly one-hour public affairs program, with film segments, call-ins, panels, interviews, etc. A biweekly half-hour film magazine called "Cabbages & Kings" features minidocumentaries, spoofs on politics, cookery, problems of students, even readings of local poetry with slides and film.

New Jersey state network ("Jerseyvision") (WNJT, Channel 52, Trenton), for innovative service to a state that has long been overshadowed by nearby metropolitan centers. In 1967 the governor of New Jersey established a blue-ribbon commission to see what could be done about serving unfilled needs for using television as a local communication medium. The result is Jerseyvision, a four-station network which became complete in 1973 with the addition of two channels in the northern part of the state. With other new educational stations, Jerseyvision shares the advantage of complete colorization from the first and the disadvantage that all its channels are in the UHF range. For the area near Philadelphia, a UHF market, this is not a problem; for the New York City area it is because many viewers are unfamiliar with UHF channels. To interest viewers in turning to UHF and to serve the state, Jerseyvision has gone into extensive local programming. ("We have more local programming than any other educational television broadcaster," a staff member told us.) The mainstay is a week-day half-hour of New Jersey news. Other highlights include weekly half-hour documentaries on state issues and a "The Editors" series where four editors from state newspapers discuss New Jersey problems that make their editorial pages and take calls from viewers. There is a black culture series and a Puerto Rican series in English and Spanish on alternate weeks. Other features include a public access program called "Potpourri," a summer magazine series called "Sunnyside Up," and a "First Person" series about prominent natives of the state. Jerseyvision will cover anything of statewide interest from a school music festival to a beauty contest, and there is considerable programming prepared by a full-time sports department. There is extensive elementary and high-school instructional television, but no college instructional television yet because a state advisory committee has yet to come up with a plan, as of 1973.

The Maryland Center for Public Broadcasting (WMPB, Channel 67, Baltimore), for innovative approaches to educational television. Maryland has an embryonic state network which will be similar in many respects to the New Jersey network, eventually with seven UHF transmitters. Two were on the air in the summer of 1973, and plans had been made for two more. The Maryland Center, in suburban Baltimore, was constructed in 1969 and already contributes extensively to the program schedules of the Eastern Educational Television Network and PBS. Notable is its "College of the Air," which functions in cooperation with several state and community colleges. In 1972–73 this college offered three credit courses per semester, reaching an average enrollment of 250 students and 300 auditors. The college also programs an ambitious "Summer Seminar," which includes a potpourri of instructional videotapes selected for their interest and timeliness. The center's basic approach to educational television, as expressed by its director, is nonpedantic. Its award-winning "Evening on Deafness" is an example. This program was presented not as reportage but as an approach to the kinds of problems deaf people encounter and the community resources that are available to help them. Such an approach to consumer education was carried out in the successful weekly series "Consumer Survival Kit." The philosophy prevails through the station's series of one-minute spots called "Etc." consisting of fascinating bits of useful information. The Maryland Center also cooperates with business and government organizations in presenting an array of management and performance training courses. There are more than twenty participating organizations.

The South Carolina State Network (WRLK, Channel 35, Columbia), for innovations in making instructional television available to the state's schools. For distributing educational materials, South Carolina's state network probably comes as close as possible to the ideal aspired to by any state network. In the state there are 250 education centers equipped for multiple-channel reception of wired programs. Including the on-the-air channel, as many as nine channels are programmed simultaneously. (The average during a school day is 3.5.) During the 1972–73 school year, 836 of the state's 1100 schools (grades 1 through 12) used instructional television on a purely voluntary basis. A total of 600,000 courses were taken by 220,000 children. In June of 1973 the state's education system graduated its first Master of Business Administration class taught entirely by television. Students watched courses at the state education centers two nights a week. According to the network, 100,000 adults have been enrolled in its professional or higher education courses. At a lower level, the network is experimenting with "Whee," a series designed for the post-"Sesame Street" age group. South Carolina produces about 75 percent of its own instructional television material.

KUAT, Channel 6, Tucson, Arizona, for innovations in the use of film and in public service programming. Film and radio, as well as television, are considered important parts of the KUAT communications package. It has a 50,000-watt daytime AM radio station, said to be "the most powerful noncommercial radio station in the USA." The film operation has won twelve Cine "Golden Eagle" awards, and its films are distributed through a number of outside sources, such as Coronet Instructional Films and the United States Information Agency. Though university owned, the TV and radio stations operate basically as community services. There is a local Spanish bilingual program for preschool children, a half-hour nightly news-in-depth show, frequent gavel-to-gavel coverage of meetings of the city council and other governmental units. Radio programs include an award-winning Alcoholics Anonymous meeting on the air with telephone participation, and a news series during which listeners can talk directly to the newsmakers.

Educational Radio

Information from the previously cited Harris study of educational television also provides a view of the scope of educational radio coverage. This service is nominally available to 46 percent of the U.S. population. It reaches significantly large segments of the East and Midwest, but outside these areas its coverage decreases proportionately with population density.

In spite of the fact that a core of AM educational radio stations exists, the total service is predominantly FM. Therefore, it is not surprising that 82 percent of those responding to the Harris survey who are able to receive educational radio programs get the service on FM. In those areas served only by FM, only 77 percent of the respondents were equipped to receive the FM signals. Thus, the Harris study concluded that in actuality educational radio was available to just 37 percent of the population. Only 9 percent of the respondents to the Harris survey had listened to educational radio within the previous week.

Some Unique Radio Services.   To supplement the television interviews, we called the managers of ten educational radio stations to determine specific programming policies and objectives.

Three were AM stations: WBAA, KOAC, and WKAR. WBAA, Lafayette, Indiana, dropped its elementary and high-school programming in 1966 to initiate an open university of the air, reaching both Purdue students and nonstudents. Since 1969, twelve different courses have been offered for credit. Enrollment reached 885 in the spring of 1973. KOAC, Corvallis, Oregon, supplies public affairs and music programming in conjunction with an FM station (KOAP) in Portland. On its subcarrier the FM station recently began a three-hour daily "Golden Age Radio" service for the elderly; the series consists largely of recordings of programs from the old days of commercial radio. WKAR, East Lansing, Michigan, recently began extensive programming services for the black audience and for migrant farm workers (largely Chicanos). The black programming occupies two and three-quarters hours each afternoon. The programming for farm workers is keyed to early and late hours in the summers. (WKAR is a daytime-only AM station.) A related FM station, on the air days and nights, programs classical music and drama (80 percent) and public affairs (20 percent).

The other stations, all FM, also have exhibited ingenuity in serving their listener groups. Powerful WAMC, Albany, New York, provides live classical music and gavel-to-gavel public affairs to a substantial section of New England. Of particular interest is its interactive communication service with health-care professionals. Carried on a subcarrier frequency, this service reaches physicians weekly in seventy hospitals and also reaches groups of nurses, dentists, medical technicians, and pharmacists. There is a half-hour lecture (illustrated by slides mailed out in advance) by a noted professional followed by telephone questions and answers. The professional may be speaking live from practically any hospital or medical school in the United States. This concept, which also has been used to a limited extent on the regular FM channel, is recognized as an important extension service by Albany Medical College.

WUOT, Knoxville, Tennessee, also is using its subcarrier frequency for medical education and is exploring use of the frequency in programming for the blind. There are two hours a month of medical programming from the University of Tennessee Memorial Hospital with talkback on WATS lines. The station also carries for-credit college extension programming and has a heavy public affairs schedule, including a weekly half hour on atomic energy produced in cooperation with Oak Ridge National Laboratory.

Probably the most successful subcarrier service for the blind is carried on through Minnesota Educational Radio, Inc., a network with powerful FM stations in Collegeville, St. Paul, Fargo-Moorhead, and soon Chandler and Rushford (Rochester). With the cooperation of a Duluth station, the network offers statewide programs for the blind seventeen hours daily on the subcarrier frequency. Most programs originate from the offices of the Minnesota State Services for the Blind in St. Paul and consist of in-depth reading of daily newspapers, news, weather bulletins, articles of special interest, and literature. A late-evening program, "For Men and Some Women," sometimes features adult novels. Presently 3200 receivers for this special service have been placed, and there are demands for an additional 5000. There also is service to adjoining states.

In East Lansing, WKAR-FM was scheduled to begin a subcarrier service for the blind and handicapped in the 1973 fall. The service has been assured of $200,000 in funding from private sources. One thousand special receivers were on order, more to be ordered soon; the ultimate potential is 15,000 receivers. Initially the service will be on the air eight hours per day, five days per week.

Milwaukee's WYMS-FM is a brand new, primarily instructional station operated by a school system that already transmits instructional television to secondary schools over two ITFS channels. (Primary grades are served by on-the-air instructional television from Milwaukee Channel 10.) The station manager believes WYMS-FM is the first station to bring total automatic programming to reel-to-reel use of instructional audio tape. The station built its own automatic programmer, adapting commercial hardware. All 156 school buildings in the Milwaukee public system have subcarrier radio service, and the station has started a faculty meeting series utilizing the subcarrier and telephone talkback.

Specifically designed to serve the needs of a predominantly black school community, the programming of WBGO-FM in Newark, New Jersey, also is directed toward serving the needs of other ethnic and cultural groups of the community. This award-winning pioneer FM station began its minority programming ten years ago. Some segments are produced by a black nationalist group, the African Free School. There are special programs in Spanish, featuring stories and civic information, and on the cultural heritage of Italian-Americans.

WMUK-FM, Kalamazoo, Michigan, recently received a grant to build a remote-control 50,000-watt transmitter. The station operates as a news, music, and public affairs medium and has won awards for its community-service programming. The station recently started monthly specials under the title of "Project '73" consisting of one-and-a-half-hour multifaceted programs on local issues with documentary, call-in, and town meeting segments. Music programming, in stereo, features live special events, such as the Ann Arbor jazz festival.

KUFM-FM in Missoula, Montana has erected a 7500-watt automated transmitter on a high mountain peak, making it by far the most powerful educational radio station in Montana or Wyoming. It was built around the philosophy of utilizing university faculty members part time, which unfortunately did not qualify it for a community service grant until it received special supplemental funding.

Funding and Administration

Noncommercial Educational Television

In its planning study, the Carnegie Commission for Educational Television anticipated a total capital investment requirement of $621 million to build a system capable of widespread local program production and local control of the program schedule. The operating expenses in support of this investment were projected at $270 million each year. Obviously, these funding goals have not been met, nor, in fact, has the funding come near the total projected by the commission.

In the eleven years between 1961 and 1972, over $252 million as capital was invested in television (see Appendix D), about 40 percent of the total proposed by the Carnegie Commission. Operating expenses reached a level of $163.9 million for the fiscal year 1972, about 60 percent of the annual level proposed by the commission. These figures have no built-in buffer against inflation. Table 3-13 compares stations' revenues and expenses for 1966-72. (However, if inflationary adjustments are made, the expenditures of average station operating funds for educational television broadcasting actually *dropped* between 1971 and 1972.) From 1966 through 1972, according to data from the National Association of Educational Broadcasters, only in one year—the first year of that period—has there been an excess of operating income over expenditures. (See figure 3-8 and table 3-13.) In 1967 and 1969 the deficit was close to $8 million each year. However, in 1972 the shortage was less than $1 million.

The gap between income and operating expense is a fact of life that each

Table 3–13
Comparison of Station Revenue to Expense—Educational Television Stations, 1966–72 ($ in thousands)

	1966	1967	1968	1969	1970	1971	1972
Number of Stations	113	119	146	189	195	207	233
Total Station Revenue	$58,315	$54,324	$66,719	$84,928	$103,641	$141,982	$162,510
Mean per Station	516	457	457	449	530	686	697
Total Station Expense	57,492	62,238	67,091	96,938	107,228	142,838	163,894
Mean per Station	509	523	460	513	550	690	703

Sources: W. Schramm and L. Nelson, *The Financing of Public Television*, Aspen Institute Program on Communications and Society, 1972.
Financial Statistics of Public Television Licensees, Fiscal Year Ending June 30, 1972. Corporation for Public Broadcasting (see Appendix D).

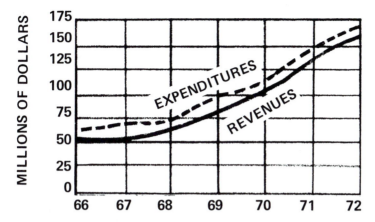

Figure 3–8. Educational Television Expenditures Compared to Revenues—1966 to 1972.

station manager must face at budget time. What does he do about it? He looks for innovative approaches to fund raising such as the annual auction, which has been a lifesaver for many stations. Sometimes he makes do with volunteer help or unpaid student help, or even with contributions in kind from a local commercial station. All too often the funding gap results in a cutback of staff or of services and a reduction of hours on the air, which it is hoped is temporary. Staff members may be asked to forego an expected salary increase. The overall result is a deterioration in the services a station provides at a time when good services are needed to attract local funding.

Of the local television broadcast organizations we contacted, twenty-seven participate in fund-raising activities, and twenty-three do not. Those who do not receive the bulk of their operating support from state education agencies or from local school districts. Often they are proscribed from fund raising by charter or statute. Yet the picture is by no means uniform. A number of stations are supported both by private and public contributions.

Twenty-nine of the TV operations have active viewer organizations of some sort; twenty-one do not, either because of lack of interest or because their charters prohibit such activities. The term "friends of public television" does not mean the same thing in all station environments. Generally, but not always, "friends" are *not* members of the public who write in for program schedules or make contributions. They often are hand-picked advisors, chosen for their civic awareness and/or ability to make relatively large contributions to a station's resources. Sometimes the "friends" are volunteer workers who take part in the legwork necessary to public television, such as office work, producing and mailing the program guide, and running errands for a TV auction and other fund-raising activities. In a few cases the "friends" are simply interested viewers who get the program guide and make minor contributions to a station's support. Some stations have separate "friends" and viewer organizations. Rather than contribute cash, local organizations sometimes contribute or loan equipment, such as a mobile van.

In 1970, the number of noncommercial educational television stations was about 28 percent of the number of commercial television stations, but the revenues constituted only about 6 percent of commercial station revenues. One reason sometimes given for this difference is that commercial stations need more television personnel. "This simply isn't so!" one station manager told us. "Our staff needs are greater than those of the commercial station (a network affiliate) because we produce more local programs."

Salary support to obtain and retain talented people was mentioned frequently by station managers as a primary need. A number of stations had hired black or Chicano producer-directors for the durations of specific programming grants; when the grants ran out, the personnel left. The funding crunch often leaves stations without development directors and public relations personnel, duties which then eat into the time of producers or managers. One station manager commented that for two years he has been so burdened with report writing and fund raising that he has lost track of his own programming. A number of stations have plans to expand their air time but lack funds for more engineers.

Table 3-14 summarizes the sources of income of educational television licensees for fiscal year 1972.

There is a tendency for state boards of education to assume a larger and larger share of ITV funding, as well as programming. However, there are many variations in ITV support, and the only conclusion that can be based on the interviews we conducted is that local fund raising is rarely if ever used to support ITV. In the fifty organizations we contacted, twenty-three get their ITV financing directly from state agencies. Ten receive funding based to some extent on a per-pupil arrangement. Several others are funded by combinations of state and local educational agencies.

It is probable that many educational television stations would not be on the air without federal support. However, because of uncertainties over the level of such support, some station managers we interviewed are reluctant to embark on long-range development plans. Since 1966, six funding plans have been devised by the NAEB (1966, 1970), the Carnegie Commission (1967), HEW/USOE (1969), and CPB (1970' 1972). These plans have called for total federal funds ranging from $155 million over ten years to $540 million over five years. The most recent authorization, for $110 million over two years, while signed by the president, still had not received funding appropriation as of the late summer of 1973. If appropriated, it will be the largest annual amount of federal support yet given to educational broadcasting, but still well below the level specified as

Table 3-14

Sources of Income—Educational Television Licensees, Fiscal 1972

	Amount ($ in thousands)	Percent	Percent Change from FY 1971
Intra-Industry Sources	$ 16,666	10.3	13.0
Federal Government	15,465	9.5	56.4
Local Schools and Boards of Education	19,379	11.9	13.7
Other Local Government Sources	3,513	2.2	13.1
State Boards of Education	14,269	8.8	-4.9
Other State Government Sources	23,926	14.7	-24.5
State Universities	18,992	11.7	114.8
Other Colleges and Universities	547	0.3	-15.4
Underwriting	4,141	2.5	25.7
National Foundations	15,904	9.8	9.6
Other Foundations	2,585	1.6	82.4
Auctions	5,672	3.5	46.1
Subscribers and Individuals	11,601	7.1	37.3
Business and Industry	2,760	1.7	-19.0
Other Production Contracts	3,524	2.2	16.2
All Other Funds	3,566	2.2	17.8
Total	$162,510	100.0	14.5

Source: Financial Statistics of Public Television Licensees, Fiscal Year Ending June 30, 1972. Corporation for Public Broadcasting (see Appendix D).

necessary in repeated projections made by the industry. The funding goal is to achieve "adequate" levels of support on a long-term basis. Such support would insulate the industry from changes in attitude toward public broadcasting prompted by political expediencies.

Fund scarcities have their inevitable effects in lack of staff or equipment. More than half of the stations we contacted do not have enough modern video-tape equipment to support production and to achieve flexibility in scheduling by delaying feeds from PBS or other sources. In only a few cases are there plans for capital investment or replacement of obsolete equipment. Both large and small stations are in a perpetual state of scrambling just to stay on the air; with obsolete equipment, maintenance cuts deeply into available staff time.

Noncommercial Educational Radio

There is some indirect support for educational broadcasting that is not recorded in the statistics. This includes the hours of time spent by volunteer helpers who assist by performing stenographic services, answering the telephone, duplicating program guides, etc. In radio, this situation is compounded by the fact that there is no statistical breakdown of revenues for those stations that do

Table 3-15
Revenues of Qualifying Educational Radio Stations by Source, Fiscal 1972

	Amount ($ in thousands)	Percent	Percent Change from FY 1971
Intra-Industry Sources	$ 1,608	10.4	62.3
Federal Government	964	6.3	45.3
Local Schools and Boards of Education	1,155	7.5	128.9
Other Local Government Sources	1,130	7.3	17.8
State Boards of Education	924	6.0	12.6
Other State Government Sources	439	2.9	-66.2
State Universities	6,464	42.0	49.7
Other Colleges and Universities	837	5.4	-18.1
Underwriting	73	0.5	-27.9
National Foundations	39	0.3	600.9
Other Foundations	182	1.2	18.0
Auctions	(a)	(a)	-98.1
Subscribers and Individuals	1,327	8.6	29.4
Business and Industry	31	0.2	-35.9
Other Production Contracts	34	0.2	358.1
Profits, Interest, and Misc.	112	0.7	27.2
All Other Funds	84	0.5	2.0
Total	$15,403	100.0	27.0

(a) Greater than 0 but less than $1000 or 0.05 percent.

Source: Public Radio Financial Statistics, Fiscal Year Ending June 30, 1972, Corporation for Public Broadcasting (see Appendix E).

not qualify for community service grants. Of the 121 radio stations that *did* qualify for such assistance at the end of fiscal year 1972, the total income for the fiscal year was $15.4 million, an increase of 27 percent over the previous year. State universities were by far the largest funding service. For these stations, a detailed breakdown of revenues by source is shown in table 3-15.

In general, the same funding inadequacies that affect the financing of non-commercial educational television also affect radio. In addition, the operators of a number of radio facilities in the Intermountain West feel that criteria for educational radio programming assistance are discriminatory because of full-time staffing requirements. It is their contention that, with automation and a minimum of staff, educational radio can provide a needed service in areas of low population density.

Conclusion

There is universal agreement among the educational broadcast organizations we contacted that the goals for funding as set forth in general by the Carnegie Commission are realistic, necessary, and attainable. However, from available statistics and from our own contacts with broadcast organizations, it is obvious that much needs to be done to provide adequate means of reaching these goals. It seems unfortunate that the search for financial support, and efforts to "make do" when support is lacking, consume a large portion of the time and talent of station staff members.

4 The Future—Alternatives and Recommendations

Educational broadcasting has passed through an initial phase of establishment and growth. To consider its future potential and alternatives, we must reexamine the goals and recognize areas of inherent ambiguity and conflict.

We must, for example, recognize that the current combination of *instructional* broadcasting service with public *cultural* broadcasting is largely a marriage of convenience. Over the next few years it seems likely that instructional technology will continue to develop and that broadcasting will be only one of the distribution alternatives. At the same time the role of public broadcasting as a cultural and educational milieu will continue to mature and complement the world of formal instruction. Yet the two represent different orientations, and it is enlightening to separate their goals.

The Goals of Public Broadcasting

Public broadcasting provides an alternative to the commercially determined environment of conventional broadcasting. It need not serve all of the people all of the time, but it surely should provide significant service to a substantial portion of the populace. To fill this role, its distribution pattern must reach most of the population with a standard of technical and production quality comparable to that of commercial broadcasting.

To provide cultural alternatives that serve a significant portion of the population, public broadcasting must be a catalyst and home for innovation and experimentation. It must constantly balance the proportion of local and national programs. And it must weigh its success not by the attraction of a single mass audience, but rather by consistent service to a variety of modest audiences that collectively comprise most of the populace.

The Goals of Instructional Broadcasting

In contrast to this overall picture of public broadcasting, we have defined instructional broadcasting as that portion in service to the formal educational community. The nature of this role has not yet been well defined or accepted, but it is surely one of service and can be weighed by this standard.

The role implies a flexible schedule adapted to the individual classroom and teacher. Thus, the distribution system should be designed to provide the material to any classroom, in a manner to suit the style and schedule of the instructor. The variety of materials, schedules, and programs is in sharp conflict with the single-channel nature of broadcasting, and if ITV were more popular, broadcast capacity would surely be overtaxed. This limitation is reflected in the current pattern of ITV . . . skewed sharply toward the primary grades, where the classroom can be more easily adapted to the medium.

Meanwhile there is a significant trend in secondary and advanced education toward adoption of "open classroom" and "open university" techniques. Every school of higher education is faced with ultimate problems of bricks and mortar—limited classrooms and limited facilities. The search for alternatives is unavoidable, and media promise a new flexibility for both instructor and student.

Distribution

In practical terms, the educational broadcasting system must be designed to reach *most* of the population rather than *all*. Commercial broadcasting, with its far greater resources, does not cover all the population, and even the most alluring new technology does not show promise of achieving universal coverage in the reasonably near future. Still, the system can be no more effective than its distribution mechanism, and new technology will probably extend the range as well as the quality and flexibility of many broadcasters.

Satellites promise a "great antenna in the sky" reminiscent of the abortive airborne experiments of early ETV. The cost of a satellite is formidable, its lifetime short, and for the foreseeable future its technical challenge is more intriguing than its probable single-channel impact on education. Not only do such experiments emphasize the cost and complexity of reaching low-population-density areas, but they underscore the inherent problems of single-channel education and the doubtful political implications of "education from above."

On the other hand, satellites may prove to be a cost-effective means of interconnecting widely dispersed local systems. A number of cable TV operators have already formed a consortium to explore this potential. It may be appropriate for the educational community to engage in research to explore the potential of direct satellite broadcasting, but it seems unnecessary to lead the way in developing a new broadband relay technique when there is more than enough commercial interest in this development.

Aside from questions of cost and technical feasibility, direct satellite transmission implies an inherent separation of the national programming and distribution pattern from that of local programming. If successful, the effect on local broadcasting would be extreme. A lack of concern about such an eventuality by local broadcasters who we interviewed probably reflects their evaluation of its feasibility. These broadcasters candidly admit that national programming is their primary asset for attracting audiences.

Broadband Transmission

Broadband interconnection systems are developing rapidly. In recent years a number of companies have been formed to provide specialized carrier service. The principal impetus for this development is the growing traffic in data communication and the need for high-speed, error-free transmission. The Bell System is matching this thrust with rapid development of its own specialized distribution system. In both cases the probable effect will be to lower the cost of broadband transmission and to make interconnection more readily available. This potential is significant to both television and radio. The current NPR distribution relies on a 5-KHz standard, which, while adequate for voice, is well below the quality needed for high-fidelity FM stereo transmission.

Broadcast distribution technology is developing rapidly to support the services and traffic we have just discussed. In effect, the growing volume of data communications, video business conferences, "cashless society" credit transactions, etc. are helping to support technology that will both make interconnection less costly and provide alternatives to open broadcasting. In perhaps ten to fifteen years, we should begin to see bundles of fiber optics supplement, or even replace, much of the current electronic distribution system.

As this happens, it is quite possible that the cost of broadband transmission will be no greater than narrowband, limited only by the cost of terminal processing. However, neither this nor satellite interconnection is close at hand, and the best that can be expected from new interconnection technology is a gradual increase in facility, combined with a gradual reduction in cost.

Emerging Technology for Distribution

Perhaps most important to the pattern of educational broadcasting is a growing potential to expand its coverage and multiply its utility through altered technology and patterns. We give this possibility high priority because adequate distribution is critical to all other goals of both public and instructional broadcasting. We have indicated that educational television is, in fact, available to only about 72 percent of the population. Educational radio, despite its low cost and adaptability to local programming, is available to far less than this percentage.

While the Intermountain West has a particularly low population density, it seems inconsistent with the aims of educational broadcasting not to provide better service for residents of this area, particularly when federal assistance is made available to stations in Alaska and American Samoa. (See the opinions of Dr. Philip Hess of the University of Montana, Appendix F.)

To further the goal of maximum coverage and competitive technical quality, we suggest that three related policies be given high priority.

1. *Radio-television Piggyback.* It seems inappropriate that most of the population is reached by educational television rather than educational radio, when the latter requires far less investment in equipment or programming. Radio is far more adaptable to local needs, such as broadcasts from remote locations. An interview with the mayor or chairman of the board of education can easily be broadcast over radio by a two-man team and a telephone line or tape-recorder. A television broadcast of the same material requires a remote truck, a crew of technicians, lighting equipment, and a major interference with the activities.

We suggest that an educational television station, in order to receive support or programming, be required to also operate an educational radio station covering the same area. This one step alone would approximately double the availability of radio, at a negligible increase in cost or inconvenience to the television broadcasters.

2. *Minimum Standards of Television Transmission Facilities.* Transmission effectiveness is prerequisite to program potential; thus, we suggest that basic broadcast capability take priority over production capability. We believe, for example, that the establishment of "slave" transmitters to extend the availability of educational broadcasting (even without capability of local origination) is preferable to no service at all.

Minimum television transmission capability should include color transmission combined with adequate power and antenna location. It also requires modern color-switching equipment and minimum of four color videotape recorders to allow a reasonable flexibility in local program schedules.

3. *Minimum Standards of Radio Transmission Facilities.* In like manner, radio FM stations receiving support should broadcast in stereo and be supported by the appropriate recording equipment. They should receive national program service and interconnection from NPR. It is appropriate that this support be based upon service to the community in terms of power and broadcast schedule, but current NPR requirements for broadcast staff are unduly restrictive. As currently administered, they eliminate support for radio facilities operated by television station personnel.

As in the case of television, additional coverage should receive high priority, and even "slave" transmission is preferable to no service at all.

Cable Television and Similar
Wired Distribution

The most striking change in distribution technology is found in the growth of cable television systems. There is little question that during the next ten years we will witness rapid growth of both the extent and technology of such systems. They are now at a stage comparable to television before the advent of commercial support–an experimental system awaiting the development of an economic base. This will probably be provided first by the advent of "pay" (subscription) television and, later, by development of ancillary transmissions such as medical services and credit transactions.

As this happens, the need for over-the-air transmission will diminish in such areas. In effect, there are already many areas where a cable operator provides a free transmitter. Yet, the Educational Broadcast Facilities Program does not provide support for organizations to capitalize on this potential. Clearly, the policy should be extended to include consideration of interconnection between current stations and cable operations, and support for facilities that produce and schedule material without operating a transmitter of any kind.

The city of Dayton, Ohio, for example, will soon have a cable network embracing the metropolitan area. Should an educational broadcast service feeding this network be considered less useful because it avoids the investment in transmitter and towers?

This same development of wired transmission foreshadows problems for the marriage of convenience between instructional and public broadcasting. This new technology offers the potential multiplicity of service and flexibility of schedule which a single over-the-air channel cannot present.

Cable system proposals for the Dayton area, for example, are asked to include consideration of a closed loop, linking schools, libraries, museums, and other institutions of learning. The goal of this dedicated network is not a new broadcast channel, but rather a communication network designed to facilitate electronic shuttling of material between locations.

With such a system, for example, it is possible to distribute class materials sequentially, to be recorded, as desired, by the local school and played back in the classroom at the time and convenience of the instructor. With such a system, the material can be interpreted, modified, and repeated, when needed, entirely free of the constraints of a broadcast schedule. Experiments like the Midwestern Educational Television CADAVRS program are leading in this direction at the same time that ITV experiments are leading toward program materials that support, rather than substitute for, the classroom. (The CADAVRS system uses switches activated by television signals to turn on and off videotape recorders located at the schools.) Such directed transmission is minimally compatible with conventional broadcasting, and we foresee an increasing difficulty of the marriage. Since most of the funding for public television derives from this activity, we anticipate both financial problems for public broadcasting and hampered growth of instructional technology, unless suitable alternatives are developed.

A variety of tools are now available for local recording and playback; some are suitable for capturing transmission during otherwise off hours. It seems appropriate to begin testing the potential use of such an approach to evaluate its utility. With such devices it is even conceivable that commercial broadcast stations might cooperate in the transmission of classroom materials during the wee hours. Most stations are quite conscious of their public-service obligation and would likely welcome an opportunity to provide public service that does not conflict with their commercial activity.

As instructional technology continues to mature and is more integrated with the curriculum, new transmission options will be critical to its success. While the separation from public broadcasting patterns may never be complete, we believe the trend is inevitable and that continued exploration of alternatives constitutes a pressing need for research.

At the same time, it is appropriate to recognize an ever broadening definition of the classroom, as exemplified by "Sesame Street" et all. The role of public broadcasting in establishing an educational milieu cannot be denied. Neither should we underestimate the profound impact of this milieu on the content and attitudes of formal education. And we must respect the growing concern for continuing education of professionals as well as for vocational training. Both of these call for access to the individual in his home at a time of his convenience. The new hardware for off-hour recording may well lead to such service.

Emerging Program Patterns

Second only to the establishment of adequate distribution facilities is the development of the programming system that can provide a balance of local and national options—a compromise between satisfying local interests and diversity with the economies, scale, and skill found in national distribution.

We seek a system that combines the virtues of experimentation, exploration, and specialized interest with the broad interests, competitive skill, and large effort that only national production and mass audience can justify. It is unfortunate, as a station manager noted sadly, that a few years ago one could produce an excellent and intriguing program on Japanese brushpainting that today's search for audience would forbid.

Despite the difficulty of this compromise and the unrealistically low funding for program production, public broadcasting has made impressive progress. National distribution has been established, together with formal channels for federal subsidy of program production as well as facilities development. We may yet this year, for the first time, see funds allocated at the starting level suggested by the Carnegie Commission Report. A few programs, such as "Civilization" and "The Forsythe Saga," have begun to attract the significant audiences that justify the cost of broadcasting, though it is interesting to note that both of these examples are BBC imports.

At the same time, due largely to the roughly quarter-billion-dollar investment by private foundations, public broadcasting has not only completely changed the preschool environment for millions of children, but has generated serious questions about our entire approach to education. As "Sesame Street" continues to penetrate the classroom, we have yet to resolve the question "Should education generally follow this lead, or will its tempo and style generate an impatience for the limitations of conventional education and the pedestrian tempo of the real world?"

As the use of instructional television increases and it assumes more of an integrated suppporting role, there will be an increasing need for information about this style of presentation and its effect. Program diversity and satisfaction of local needs are difficult goals to reconcile with limited resources and the inherently costly investment in television programming. One must recognize that diversity rather than uniformity is the predominant character of public broadcasting today, and that it represents a collection of individual broadcasters reflecting a variety of motives and goals. Some are noble; some are petty and territorial, linked together largely through self-interest and common problems. National domination is a doubtful threat, and while there may be many local audiences and needs unserved, it is doubtful that duplication of licenses is appropriate.

To serve a necessarily limited portion of the total audience, it seems clear that good service and high technical quality should take precedence over the multiple service that has begun to appear in some areas. We see an apparent need for an effective policy for both "birth control" and standards of equipment. The limited resources available can surely provide greater service by adding color capability, modern switching gear, increased transmitter power or relay transmitters to extend the service of a current facility, rather than by building a new station to fragment the audience of an existing one.

We suggest that the basic criteria of such a policy are not difficult to formulate. The law calls for both maximum coverage and a balance of local, regional, and national production. We suggest that the U.S. Office of Education define (and fund accordingly) three levels of distribution and production facility.

1. *The Local Television Distribution Center.* In today's environment, *every* station should be capable of transmitting in color, as well as providing a signal of good strength and quality throughout its area, and altering schedules to suit local needs and desires. This implies color capability, modern switching gear, a minimum of four videotape recorders, optimum transmitter power and location, and, in some cases, relay transmitters or other devices to extend coverage. (Without exception, managers of noncolor stations who we interviewed stressed the need for color to compete with commercial channels.)

This baseline transmission should receive high priority, and no station can be considered truly adequate unless it meets these criteria. At the present time, many do not.

As discussed previously, we believe that this goal of standard minimum transmission capability, extended to the maximum reasonable audience through cable systems and similar techniques as they become available, should receive the highest priority, even at the expense of establishing "slave" facilities with little or no production capability.

2. *Local Production Facilities.* We suggest a study to establish a standard of minimum equipment suitable for local program production. The goal of this study would be a general standard for facility and equipment, adequate to complement the PBS feed and to provide the capability of originating a reasonable variety of programming. Yet, the resources should be distributed to avoid the current disparity between "have" and "have-not" stations.

If the goal is broad service to all communities, it is an incongruity to apply federal funding to equip one station with multiple studios and idle cameras, while another station modifies obsolete black-and-white switching hardware simply to survive. While the actual definition of standards for such a facility will require substantial study, a brief and tentative discussion might suggest the kind of facility to be envisaged.

Such a standard should consider the current range of local program activity and the facilities that have developed to meet this need, but it should also consider the facilities of commercial counterparts who are nothing if not conscious of the cost-benefit tradeoffs of local programming.

This capability might include a basic contingent of three color cameras and an appropriate control room attached to a studio of significant physical dimensions. Limited studio space represents a poor economy. A number of sets can be arranged within a large area, eliminating the logistics and inconvenience of constant setup and teardown. At the same time, camera and lighting equipment can serve multiple duty, minimizing both investment and maintenance.

A minimal local facility should also include sufficient supporting hardware: the four color videotape recorders mentioned previously, as well as appropriate color equipment for editing, and a film chain. It is doubtful whether mobile television equipment represents an appropriate part of a minimal facility. The investment in equipment, maintenance, and operating crew make mobile operation one of the most difficult activities, and it is for this reason that so few commercial stations own such equipment. We know of one commercial station that owns a well-equipped color mobile unit, but places severe restrictions on its operation because of the high cost of its use and maintenance.

A better alternative for local production would be an adequate film capability—at least one or two single-system cameras and professional editing equipment. The details of such a "standard facility" should be the subject of serious study, and there may be reason for variation to meet local needs. But the need for such guidelines in allocating funds seems clear. This will be especially urgent if, as we anticipate, the growth of instructional programming causes a proliferation of production activity.

While it is appropriate to equip all basic distribution facilities with this level of production capability, ownership of a transmitter should not be a prerequisite. And it seems appropriate that the Part IV of Title III of the Communications Act of 1934, as amended, should be modified to allow both for program production that will be transmitted through a cable or similar facility, and to allow the development of production centers that can supply instructional needs through other distribution mechanisms.

3. *Regional Production Centers.* It is not possible at reasonable cost to equip or operate all (or even most) stations above the level just described. It would indeed be a significant accomplishment if all stations could be maintained at this minimal level. Such facilities would provide a training ground and a base for experimentation, as well as answering the need for local program production. In the aggregate, they could provide a channel for identifying, developing, proving, and encouraging promising new talent.

However, more is needed, both in capability and potential, to compete with the skill and flamboyance of heavily capitalized commercial broadcasters. This implies the need for large production centers along the lines of those now feeding PBS, facilities characterized by extensive capabilities such as specialized equipment, but more importantly, skilled professional technicians and camera crews, with experience, talent, and financial support.

We suggest a deliberate policy to fund a number of such regional centers, geographically distributed to provide both regional service and national input. These facilities should be administered as regional assets, controlled and operated by consortia. Thus, a mobile unit might travel from place to place, serving the constituent stations as a central resource. Such a facility might be more likely to engage in major projects than to assemble an empire of idle equipment.

We recognize the political difficulties of these suggestions and the deliberate allocation of funds we have implied, but they reflect our view of problems that must be faced if educational broadcasting is to continue to make significant progress toward its ambitious goals.

Within these guidelines, we believe cost-sharing and matching-fund criteria are useful means to assure responsiveness to local needs. But we believe that there is an additional need for improved techniques to report and compare activity. We see a great need for a management reporting system that can identify and compare the general activities and levels of effort in individual stations, and, perhaps more important, identify common problems of obsolescence, inadequate capabilities, funding, and administration.

In a recent study of audiovisual activity management for a major government agency, we developed a reporting system that graphically pointed out exceptionally inadequate equipment, staff, and activity. An effort of this type requires a systematic analysis of the activities to be meaningful, as well as ingenuity in devising parameters that can be measured and reported without excessive difficulty. But with skill, a system can be developed and made to provide useful information, both for the participants and for those monitoring the activity.

5 Summary

We foresee a gradual but unavoidable separation between public broadcasting and the support of formal instruction. We are inclined to agree with the recent forecast of the Bell System of Canada, that within school buildings, internal cable distribution of television is inevitable in the reasonably near future.[1] We believe further that such facilities will be supplied through some type of broadcast distribution system, either single-channel sequentially, like the current Minnesota system, or more likely through some variety of multi-channel cable distribution system.

This may well be accomplished in a dedicated facility, linking the components of a school system together with other cultural centers such as museums and libraries. The end result will be an electronic distribution system in support of formal education, coupled with a variety of terminal recording equipment designed to "catch" this transmission and make it available at the convenience of the instructor. This picture does not fit the current pattern of the Educational Broadcasting Facilities Program, and we suggest that its interpretation be broadened to include this potential.

Public broadcasting, on the other hand, will probably continue to mature and develop a character of its own as the alternative to commercial broadcasting. One must recognize that the goal of a high-quality alternative to commercial broadcasting, providing both a cultural and informal educational milieu, is intangible and may never truly be met. It will always be an art to balance programs of limited audience appeal against those with broad appeal. And major program success in the public sector will probably tend to move to commercial broadcasting, either in fact or through imitation. The effects of "Sesame Street" on children's daytime programming already are apparent. So are the effects of "The Forsythe Saga" on adult drama.

Television production is inherently extremely expensive, requiring the skill and artistry of large teams supported by heavy capital investment. Its cost can be tolerated only because of the enormous audience. As a result, we place first priority on the continued development of the distribution network, allocating available resources to minimum standards of transmission equipment and such capabilities as color transmission, adequate videotape recorders, and "slave" transmitters.

We have suggested that programming facilities be considered separately, and that the balance between local programming and national service can best be serviced by defining three levels of operation and equipment and facilities.

1. Transmission only, using both locally and nationally generated materials.

2. Local program generation comprising a modest color origination facility, together with editing equipment and limited film production tools.

3. Regional program centers incorporating the extensive facilities and tools (such as mobile units) needed to compete with the scale and flamboyance of commercial broadcasting. These facilities would be regional in role as well as location, operated by and in the service of a regional consortium or network. They would also serve as important sources of programming for national distribution.

While transmitters are at present the core of educational broadcasting, there is already a trend toward alternative distribution systems such as cable television. Accordingly, the EBFP interpretation should be broadened to include program production facilities, established for the purpose of large-scale distribution, but not necessarily attached to a transmitter.

We have suggested a number of areas that urgently call for further study. There is a need to establish guidelines of equipment performance and obsolescence that can grow with state-of-the-art developments in the three categories that we suggest. There is also a need for better monitoring of activity. We have suggested development of a management-reporting system that can reflect problems and inadequacies in the public broadcasting system. We believe it is possible to collect more useful measures of the actual service provided by public broadcasting than those furnished through statistics alone.

We have tended to focus attention on television because of its larger investment and more complex technical and program problems, but we have attempted to parallel this with discussion of educational radio. And we are concerned lest the dominant television investment overshadow the valuable contribution of radio. The limited "coverage" of educational radio should not continue, and we have suggested a requirement that each ETV station also operate a radio facility as a simple, inexpensive means to greatly extend radio service. We have suggested minimum radio facility standards to include high-fidelity stereo transmission; the requirements that currently prohibit support to stations with part-time or shared personnel should also be revised.

The potential of educational broadcasting has been demonstrated, but is by no means fulfilled. Both instructional and public broadcasting deserve significantly greater support, and it is difficult to reconcile the ambitious goals with the support currently available. Yet, the limited resources must be husbanded and allocated to produce the maximum impact. As long as this is so, primary emphasis must be focused on extended and improved transmission, and the funding of production capabilities must be carefully balanced between the need for substantial pools of talent and equipment and the desire for local activity.

Telephone Survey of Television Stations

Local Educational Agencies

Miami, Florida: WTHS Channel 2, WSEC Channel 17
Atlanta, Georgia: WETV Channel 30
Las Vegas, Nevada: KLVX Channel 10
Oklahoma City, Oklahoma: KOKH Channel 25, KETA Channel 13
Nashville, Tennessee: WDCN Channel 2
Yakima, Washington: KYVE Channel 47

Colleges and Universities

Orono, Maine: WMEB Channel 12, also WMED Channel 13 Calais and WMEM
    Channel 10 Presque Isle
Durham, New Hampshire: WENH Channel 11, also WEDB Channel 40 Berlin,
    WHED Channel 15 Hanover, WEKW Channel 52 Keene, WLED Channel 49
    Littleton
Burlington, Vermont: WETK Channel 33, WVER Channel 28 Rutland, WVTB
    Channel 20 St. Johnsbury, WVTA Channel 41 Windsor
Austin-San Antonio, Texas: KLRN Channel 9
Las Cruces, New Mexico: KRWG Channel 22
Salt Lake City, Utah: KUED Channel 7
Pocatello, Idaho: KBGL Channel 10
Brookings, South Dakota: KESD Channel 8
Pueblo-Colorado Springs, Colorado: KTSC Channel 8
Topeka, Kansas: KTWU Channel 11
Tucson, Arizona: KUAT Channel 6
Albuquerque, New Mexico: KNME Channel 5
Chapel Hill, North Carolina: WUNC Channel 4, also WUNF Channel 33 Ashe-
    ville, WTVI Channel 42 Charlotte, WUND Channel 2 Columbia, WUNG
    Channel 58 Concord, WUNK Channel 25 Greenville, WUNE Channel 17
    Linville, WUNJ Channel 29 Wilmington, WUNL Channel 26 Winston-Salem

College Station, Texas: KAMU Channel 15
Carbondale, Illinois: WSIU Channel 8

State Agencies

Columbia, South Carolina: WRLK Channel 35, also WEBA Channel 14 Allendale,
    WITV Channel 7 Charleston, WJPM Channel 33 Florence, WNTV Chan-
    nel 29 Greenville
Birmingham, Alabama: WBIQ Channel 10, also WIIQ Channel 41 Demopolis,
    WDIQ Channel 2 Dozier, WFIQ Channel 36 Florence, WHIQ Channel 25
    Huntsville, WGIQ Channel 43 Louisville, WEIQ Channel 42 Mobile,
    WAIQ Channel 26 Montgomery, WCIQ Channel 7 Mount Cheaha
Jackson, Mississippi: WMAA Channel 29, also WMAB Channel 2 Ackerman,
    Channel 12 Booneville (not yet on the air), WMAU Channel 17 Bude,
    WMAO Channel 23 Inverness, WMAH Channel 19 McHenry, WMAV
    Channel 18 Oxford, WMAW Channel 14 Rose Hill
Lexington, Kentucky: WKLE Channel 46, also WKAS Channel 25 Ashland,
    WKGB Channel 53 Bowling Green, WCVN Channel 54 Covington, WKZT
    Channel 23 Elizabethtown, WKHA Channel 35 Hazard, WKMJ Channel 68
    Louisville, WKMA Channel 35 Madisonville, WKMR Channel 38 More-
    head, WKMU Channel 21 Murray, WKON Channel 52 Owenton, WKPI
    Channel 22 Pikeville, WKSO Channel 29 Somerset
Beckley, West Virginia: WSWP Channel 9
Little Rock, Arkansas: KETS Channel 2
Tulsa, Oklahoma: KOED Channel 11
Des Moines, Iowa: KDIN Channel 11, also KIIN Channel 12 Iowa City
Corvallis, Oregon: KOAC Channel 7, also KOAP Channel 10 Portland
Rapid City, South Dakota: KBHE Channel 9, also KDSD Channel 16 Aberdeen,
    Channel 13 Eagle Butte, KTSD Channel 10 Pierre, KUSD Channel 2
    Vermillion
Chatsworth, Georgia: WCLP Channel 18, also WDCO Channel 15 Cochran,
    WJSP Channel 28 Columbus, WACS Channel 25 Dawson, WABW Chan-
    nel 14 Pelham, WVAN Channel 9 Savannah, WXGA Channel 8 Waycross,
    WCES Channel 20 Wrens (Augusta)
Boise, Idaho: KAID Channel 4
Trenton, New Jersey: WNJT Channel 52, also WNJS Channel 23 Camden,
    WTLV Channel 58 New Brunswick and Channel 50 Montclair
Baltimore, Maryland: (Owings Mills) WMPB Channel 67, also WCPB Channel 28
    Salisbury

Community Stations

Syracuse, New York: WCNY Channel 24
New York City: WNYE Channel 25
Hartford, Connecticut: WEDH Channel 24, also WEDW Channel 49 Bridgeport,
    WEDN Channel 53 Norwich
Norfolk, Virginia: WHRO Channel 15
Roanoke, Virginia: WBRA Channel 15
Detroit, Michigan: WTVS Channel 56
St. Paul, Minnesota: KTCA Channel 2, also KTCI Channel 17
Fargo, North Dakota: KFME Channel 13 also Channel 2 Grand Forks (not
    yet on air)
New Orleans, Louisiana: WYES Channel 12
Jacksonville, Florida: WJCT Channel 7
Redding, California: KIXE Channel 9

State Agency by Letter

Pago Pago, American Samoa: KVZK Channel 2

Telephone Survey of Radio Only Stations

Albany, New York: WAMC-FM
Knoxville, Tennessee: WUOT-FM
Kalamazoo, Michigan: WMUK-FM
Newark, New Jersey: WBGO-FM
Milwaukee, Wisconsin: WYMS-FM
Lafayette, Indiana: WBAA-AM
East Lansing, Michigan: WKAR-AM and FM
Collegeville, Minnesota: KSJR-FM
Missoula, Montana: KUFM-FM
Minneapolis, Minnesota: KUOM-AM

Stations Visited

Boston, Massachusetts: WGBH
Pittston, Pennsylvania: WVIA

Washington, D.C.: WETA
Pittsburgh, Pennsylvania: WQED
Chicago, Illinois: WTTW, WXXW
Los Angeles, California: KCET
San Francisco, California: KQED
Bloomington, Indiana: WTIU
Lincoln, Nebraska (total of nine stations on Nebraska network) KUON
Columbus, Ohio: WOSU

By Call Letters

KAID
    Boise, ID
KAMU
    College Station, TX
KBGL
    Pocatello, ID
KBHE
    Rapid City, SD
KCET
    Los Angeles, CA
KDIN
    Des Moines, IA
KDSD
    Aberdeen, SD
KESD
    Brookings, SD
KETA
    Oklahoma City, OK
KETS
    Little Rock, AR
KFME
    Fargo, ND
KIIN
    Iowa City, IA
KIXE
    Redding, CA

KLRN
   Austin-San Antonio, TX
KLVX
   Las Vegas, NV
KNME
   Albuquerque, NM
KOAC
   Corvallis, OR
KOAP
   Portland, OR
KOED
   Tulsa, OK
KOKH
   Oklahoma City, OK
KQED
   San Francisco, CA
KRWG
   Las Cruces, NM
KSJR-FM
   Collegeville, MN
KTCA
   St. Paul, MN
KTCI
   St. Paul, MN
KTSC
   Pueblo-Colorado Springs, CO
KTSD
   Pierre, SD
KTWU
   Topeka, KS
KUAT
   Tucson, AZ
KUED
   Salt Lake City, UT
KUFM-FM
   Missoula, MT
KUOM-AM
   Minneapolis, MN
KUON
   Lincoln, NB
KUSD
   Vermillion, SD

KVZK
    Pago Pago, American Samoa
KYVE
    Yakima, WA

WABW
    Pelham, GA
WACS
    Dawson, GA
WAIQ
    Montgomery, AL
WAMC-FM
    Albany, NY
WBAA-AM
    Lafayette, IN
WBGO-FM
    Newark, NJ
WBIQ
    Birmingham, AL
WBRA
    Roanoke, VA
WCES
    Wrens (Augusta), GA
WCIQ
    Mount Cheaha, AL
WCLP
    Chatsworth, GA
WCNY
    Syracuse, NY
WCPB
    Salisbury, MD
WCVN
    Covington, KY
WDCN
    Nashville, TN
WDCO
    Cochran, GA
WDIQ
    Dozier, AL
WEBA
    Allendale, SC

WEDB
    Berlin, NH
WEDH
    Hartford, CT
WEDN
    Norwich, CT
WEDW
    Bridgeport, CT
WEIQ
    Mobile, AL
WEKW
    Keene, NH
WENH
    Durham, NH
WETA
    Washington, DC
WETK
    Burlington, VT
WETV
    Atlanta, GA
WFIQ
    Florence, AL
WGBH
    Boston, MA
WGIQ
    Louisville, AL
WHED
    Hanover, NH
WHIQ
    Huntsville, AL
WHRO
    Norfolk, VA
WIIQ
    Demopolis, AL
WITV
    Charleston, SC
WJCT
    Jacksonville, FL
WJPM
    Florence, SC
WJSP
    Columbus, GA

WKAR-AM and FM
    East Lansing, MI
WKAS
    Ashland, KY
WKGB
    Bowling Green, KY
WKHA
    Hazard, KY
WKLE
    Lexington, KY
WKMA
    Madisonville, KY
WKMJ
    Louisville, KY
WKMR
    Morehead, KY
WKMU
    Murray, KY
WKON
    Owenton, KY
WKPI
    Pikeville, KY
WKSO
    Somerset, KY
WKZT
    Elizabethtown, KY
WLED
    Littleton, NH
WMAA
    Jackson, MS
WMAB
    Ackerman, MS
WMAH
    McHenry, MS
WMAD
    Inverness, MS
WMAU
    Bude, MS
WMAV
    Oxford, MS
WMAW
    Rose Hill, MS

WMEB
Orono, ME
WMED
Calais, ME
WMEM
Presque Isle, ME
WMPB
Owings Mills, MD
WMUK-FM
Kalamazoo, MI
WNJS
Camden, NJ
WNJT
Trenton, NJ
WNTV
Greenville, SC
WNYE
New York City, NY
WOSU
Columbus, OH
WQED
Pittsburgh, PA
WRLK
Columbia, SC
WSEC
Miami, FL
WSIU
Carbondale, IL
WSWP
Beckley, WV
WTHS
Miami, FL
WTIU
Bloomington, IN
WTLV
New Brunswick, NJ
WTTW
Chicago, IL
WTVI
Charlotte, NC
WTVS
Detroit, MI

WUNC
   Chapel Hill, NC
WUND
   Columbia, NC
WUNE
   Linville, NC
WUNF
   Asheville, NC
WUNG
   Concord, NC
WUNJ
   Wilmington, NC
WUNK
   Greenville, NC
WUNL
   Winston-Salem, NC
WUOT
   Knoxville, TN
WVAN
   Savannah, GA
WVER
   Rutland, VT
WVIA
   Pittston, PA
WVTA
   Windsor, VT
WVTB
   St. Johnsbury, VT
WXGA
   Waycross, GA
WXXW
   Chicago, IL
WYES
   New Orleans, LA
WYMS-FM
   Milwaukee, WI

Appendix B:
Communication Groups Contacted

Association for Instructional Television (formerly National Instructional Television)
Donald Sandberg

Central Educational Network
Raymond Giece

Corporation for Public Broadcasting
John Golden, Director Planning, Research and Evaluation
Matthew Coffey

Eastern Educational Television Network
John Porter

Eastern Public Radio Network
Albert Fredette

Educational Broadcasting Review
A. Edward Foote, Editor

Federal Communications Commission
Robert Hilliard

Ford Foundation
Tinka Nobbi

Great Plains National Instructional Television Library
Paul Schupbach

Maryland Center for Public Broadcasting
Frederick Breitenfeld, Executive Director

Midwestern Educational Television, Inc.
W.D. Donaldson

National Association of Educational Broadcasters
NAEB Educational Television Stations
H. Holt Riddleburger, Deputy Director

NAEB National Educational Radio
James Robertson, Director

National Public Radio
James Barrett, Director of Public Information

Office of Telecommunications Policy—Executive Office of The President
Henry Goldburg, General Council
Vincent Sardella

Ohio Educational Television Network Commission
Richard Hull, Chairman
Dave Fornshell, Executive Director

Public Broadcasting Service
Gerald Slater, General Manager
Robert Mott, Director of Station Relations
Pepper Weiss
David Lacy

Southern Education Network
Wayne Seal

Stanford University
Lyle Nelson

University of Minnesota
Burton Paulu

University of Montana
Philip Hess

Western Educational Network
Walter Schaar

Appendix C:
Public Television Service Minimum
Annual Operating Costs

Total Annual Minimum Operating Cost, National PTV Service

Service	Cost
1. *National Program Service*	
a) Daytime Children's Programming, Regular Season	
12 hours/week, 39 weeks/year @ $ 54,000/hour average	$ 25,272,000
b) Prime-time Programming, Regular Season	
24 hours/week, 39 weeks/year, subdivided as follows:	
– Programming for Older Children and Teens	
6 hours/week @ $ 80,000/hour average	18,720,000
– Music, Drama, Performance, & Criticism	
8 hours/week @ $ 70,000/hour average	21,840,000
– Public Affairs Programming	
6 hours/week @ $ 74,000/hour average	17,316,000
– Special Programming and Special Events Coverage	
4 hours/week @ $ 100,000/hour average	15,600,000
c) Summer Programming	
Assume primarily re-runs, but with some new public affairs production and special events coverage.	
13 weeks @ 50% regular season rate	16,458,000
[Total, National Program Service]	[115,206,000]
2. *Regional Program Services*	
To stimulate regional production.	
Assume 10 regional production operations	
2 hours/week, 39 weeks/year @ $ 25,000/hour	19,500,000
3. *Local Program Services*	
a) To maintain current services	
Total PTV Station Expenditures, Fiscal Year 1971*	142,838,000*
b) To upgrade local programming	
Assume 160 stations engaged in production	
4 hours/week, 39 weeks/year @ 5,000/hour	124,800,000
[Total, Local Program Services]	[267,638,000]
4. *Innovation Support*	
To develop new sources of programs, and to experiment in new production techniques, etc.	8,000,000
5. *Interconnection*	
Assume expansion of the current system to include new stations	12,000,000
6. *Non-broadcast Activities*	
Administration, promotion, research, etc.	9,500,000
Estimated Total Annual Minimum Operating Cost	$ 431,844,000
Less: Total Unduplicated PTV System Revenues Fiscal Year 1971	165,632,000
NET MINIMUM INCREMENT IN ANNUAL OPERATING COST	$ 266,212,000

Note: *Total PTV Station Expenditures provide a realistic *approximation* of current local programming costs. Stations reported Direct Operating Costs of $ 113,242,000 in FY 1971. The total figure used here includes some funds for capital purchases and national production to offset the fact that in few cases do stations include depreciation in Operations.

Source: W. Schramm and L. Nelson, *The Financing of Public Television,* Aspen Institute Program on Communications and Society, Palo Alto (1972).

Appendix D:
Financial Statistics for Public
Television Licensees, Fiscal
Year Ending June 30, 1972

Preliminary Report of Public Television Licensees' Income: Fiscal Years 1971 and 1972

Source of Income	Fiscal 1971		Fiscal 1972		Percent Increase or Decrease
	Amount	Percent	Amount	Percent	
Intra-Industry Sources	$ 14,745,494	10.4	$ 16,666,119	10.3	13.0
Federal Government	9,885,460	7.0	15,465,246	9.5	56.4
Local Schools and Boards of Education	17,045,077	12.0	19,378,920	11.9	13.7
Other Local Government Sources	3,106,645	2.2	3,513,114	2.2	13.1
State Boards of Education	15,011,138	10.6	14,268,803	8.8	(4.9)
Other State Government Sources	31,673,011	22.3	23,925,938	14.7	(24.5)
State Universities	8,843,535	6.2	18,992,195	11.7	114.8
Other Colleges and Universities	646,422	0.5	547,109	0.3	(15.4)
Underwriting	3,294,794	2.3	4,141,351	2.5	25.7
National Foundations	14,515,464	10.2	15,904,218	9.8	9.6
Other Foundations	1,417,439	1.0	2,585,156	1.6	82.4
Auctions	3,883,302	2.7	5,671,592	3.5	46.1
Subscribers and Individuals	8,447,569	6.0	11,600,913	7.1	37.3
Business and Industry	3,408,006	2.4	2,759,791	1.7	(19.0)
Other Production Contracts	3,032,570	2.1	3,524,006	2.2	16.2
All Other Funds	3,026,270	2.1	3,566,010	2.2	17.8
Total Income	$141,982,196	100.0	$162,510,481	100.0	14.5

Preliminary Report of Percent Increase in Public Television Licensees' Income by Type of Licensee: Fiscal Year 1972

	Percent Increase or (Decrease) Over FY 1971	Number of Licensees Reporting	
		(FY 71)	(FY 72)
University	22.9	46	51
School	4.8	21	21
State and Other	(3.5)	21	22
Community	22.7	47	51
All Licensees	14.5	135	145

Preliminary Report of Public Television Licensees' Income Attributed to ITV Services: Fiscal Year 1972

Type of Licensee	Amount	Percent of Total Income
University	$ 4,597,000	13.8
School	5,903,228	59.7
State and Other	9,650,465	26.0
Community	8,117,001	9.9
All Licensees	$28,267,694	17.4

Preliminary Report of Public Television Licensees' Direct Operating Costs: Fiscal Year 1972

Purpose of Costs		Amount	Percent of Total
Technical	- Salaries	$ 17,702,534	21.9
	- Other Costs	9,350,391	
Programming	- Salaries	9,336,005	13.9
	- Other Costs	7,848,124	
Production	- Salaries	17,631,542	27.1
	- Other Costs	15,836,291	
Instructional	- Salaries	4,584,887	6.9
	- Other Costs	3,880,579	
Development	- Salaries	1,900,265	4.3
	- Other Costs	3,401,060	
Promotion	- Salaries	1,584,105	2.7
	- Other Costs	1,785,147	
Training	- Salaries	255,597	0.8
	- Other Costs	739,196	
General and Administrative	- Salaries	10,497,480	16.7
	- Other Costs	10,126,148	
All Other	- Salaries	1,911,769	5.7
	- Other Costs	5,115,107	
Total	- Salaries	65,404,184	53.0
	- Other Costs	58,082,043	47.0
Total Direct Operating Costs		$123,486,227	100.0

Preliminary Report of Public Television Licensees' Total Expenditures: Fiscal Years 1971–72

	Fiscal 1971		Fiscal 1972		Percent Increase
	Amount	Percent	Amount	Percent	
Total Direct Operating Costs	$113,242,155	79.3	$123,486,227	75.3	9.0
Total Capital Expenditures	29,596,084	20.7	40,407,708	24.7	36.5
Total Direct Expenditures	$142,838,239	100.0	$163,893,935	100.0	14.7
Other Costs Absorbed by Supporting Institutions	7,508,981		9,407,236		25.3
Total Direct and Indirect Expenditures	$150,347,220		$173,301,171		15.3

Preliminary Report of Public Television Licensees' Capital Expenditures Fiscal Year 1972

	FY 1972 Capital Expenditures	Percent Increase or (Decrease) Over FY 1971
University	$12,213,862	56.2
School	1,240,768	(47.2)
State and Other	10,494,252	23.9
Community	16,458,826	50.2
All Licensees	$40,407,708	36.5

Total Capital Expenditures to Date: $252,204,207 (as of June 30, 1972)

Appendix E:
Financial Statistics for Public Radio Stations—
Fiscal Year Ending June 30, 1972

Preliminary Report of Public Radio Stations' Income: Fiscal Years 1971 and 1972

Source of Income	Fiscal 1971		Fiscal 1972		Percent Increase or Decrease
	Amount	Percent	Amount	Percent	
Intra-Industry Sources	$ 990,776	8.2	$ 1,607,765	10.4	62.3
Federal Government	662,763	5.5	963,869	6.3	45.3
Local Schools and Boards of Education	504,580	4.2	1,155,088	7.5	128.9
Other Local Government Sources	958,510	7.9	1,129,509	7.3	17.8
State Boards of Education	820,296	6.8	923,678	6.0	12.6
Other State Government Sources	1,299,856	10.7	439,244	2.9	(66.2)
State Universities	4,317,340	35.6	6,463,595	42.0	49.7
Other Colleges and Universities	1,021,952	8.4	836,600	5.4	(18.1)
Underwriting	101,283	0.8	72,977	0.5	(27.9)
National Foundations	5,600	*	39,249	0.3	600.9
Other Foundations	154,177	1.3	181,969	1.2	18.0
Auctions	38,262	0.3	730	*	(98.1)
Subscribers and Individuals	1,025,034	8.4	1,326,743	8.6	29.4
Business and Industry	47,864	0.4	30,681	0.2	(35.9)
Other Production Contracts	7,462	0.1	34,186	0.2	358.1
Profits, Int. & Misc. Sales and Serv.	88,437	0.7	112,481	0.7	27.2
All Other Funds	82,701	0.7	84,340	0.5	2.0
Total Income	$12,126,893	100.0	$15,402,704	100.0	27.0

*Percent more than 0 but less than 0.05.

Preliminary Report of Increase in Public Radio Income by Type of Licensee: Fiscal Year 1972

	Percent Increase Over FY 1971	Number of Stations Reporting	
		(FY 71)	*(FY 72)*
University	22.8	74	84
School	72.9	8	11
State and Other	33.1	10	9
Community	27.6	11	17
All Licensees	27.0	103	121

Preliminary Report of Public Radio Stations' Direct Operating Costs— Fiscal Year 1972

Purpose of Costs		*Amount*	*Percent of Total*
Technical	- Salaries	$ 2,663,751	24.6
	- Other Costs	498,850	
Programming	- Salaries	2,670,577	25.9
	- Other Costs	662,278	
Production	- Salaries	1,621,568	15.3
	- Other Costs	351,424	
Instructional	- Salaries	231,811	2.3
	- Other Costs	62,846	
Development	- Salaries	69,187	0.9
	- Other Costs	43,350	
Promotion	- Salaries	140,374	2.5
	- Other Costs	175,428	
Training	- Salaries	161,965	1.4
	- Other Costs	19,393	
General and Administrative	- Salaries	2,196,701	22.5
	- Other Costs	698,866	
All Other	- Salaries	375,174	4.8
	- Other Costs	237,114	
Total	- Salaries	10,131,108	78.7
	- Other Costs	2,749,549	21.3
Total Direct Operating Costs		$12,880,657	100.0

Preliminary Report of Public Radio Stations' Total Expenditures for Fiscal Years 1971 and 1972

	Fiscal 1971		Fiscal 1972		Percent Increase
	Amount	Percent	Amount	Percent	
Total Direct Operating Costs	$10,500,033	88.4	$12,880,657	81.8	22.7
Total Capital Expenditures	1,375,028	11.6	2,863,317	18.2	108.2
Total Direct Expenditures	$11,875,061	100.0	$15,743,974	100.0	32.6
Other Costs Absorbed by Supporting Institutions	$ 1,914,352		$ 2,171,370		13.4
Total Direct and Indirect Expenditures	$13,789,413		$17,915,344		29.9

Preliminary Report of Public Radio Station Capital Expenditures— Fiscal Year 1972

	FY 1972 Capital Expenditures	Percent Increase Over FY 1971
University	$1,882,472	106.4
School	78,039	174.8
State and Other	250,723	1941.4
Community	652,083	54.5
All Licensees	$2,863,317	108.2

Total Capital Expenditures to Date: $18,259,285 (as of June 30, 1972)

Appendix F:
"CPB and the Small Market Public Broadcaster"

Statement of Philip J. Hess, Ph.D., University of Montana, Before CPB Long-Range Financing Hearing, October 31, 1972

My name is Philip J. Hess. I am chairman of the Radio-Television Department at the University of Montana in Missoula, a small town on the western slope of the Rocky Mountains. Our department has operated a low-power non-commercial FM station since 1965.

We have long believed it is the responsibility of the broadcaster to do everything within his technical and economic ability to service his area.

Our economic ability is fixed by the state legislature. We have increased our technical ability with equipment purchased under an HEW facilities grant. On a clear night, with the wind behind us, our new transmitter will cover 5300 square miles, including about one-half of the Flathead Indian Reservation. Our broad signal will reach only about 57,000 people in that country, where the population density is eleven persons per square mile.[a] Those 57,000 include a large number of on- and off-reservation Indians, paper-pulp mill workers, U.S. Forest Service employees, University faculty and staff members, students, and a normal distribution of other occupations.

About 15,000 of the homes reached subscribe to the local daily newspaper, 8600 subscribe to *Readers Digest,* 4000 to *Time* and *Newsweek,* and 2200 to *Playboy.*

We who operate small market public radio stations in sparsely populated areas are often guilty of failure to communicate both our achievements, of which we are exceedingly proud, and our problems. We enjoy being involved, in depth, in activities which will benefit our neighbors and friends. We are conscious, every waking moment, of the tremendous responsibility which public broadcasting imposes upon us.

We are proud that we are the only radio station in all of Montana with two wire service printers. We are proud that we are one of the few, if not the only, radio station in the country with two news department staff members who were recipients of CBS News and Public Affairs Fellowships. We are proud of what we have been able to accomplish with a limited staff.

---

[a]That is the figure for our heavily populated portion of Montana. The statewide density is 4.6 per square mile. Idaho is 8.1, Wyoming is 3.4, and Nevada is 2.6.

The National Association of Broadcasters has a small market committee. (They kindly call it "Secondary Market Committee"—that sounds better to advertisers.) One of the founding members and a former chairman of that group, is Dale G. Moore. Significantly, Dale is a Missoula broadcaster. He reports that station staffs in our part of the country average *seven persons*. And note, please, that he is giving figures for commercial stations.

Deduct from that number, if you will, those staff members whose jobs are related to the commercial operation of the station and you"ll realize that almost no commercial station in Montana, Idaho, Wyoming, or Nevada would meet CPB's full-time staff minimum of three by January, let alone five by 1976.

Ladies and gentlemen, it is not by accident or oversight that there is no noncommercial, public radio station in that four-state area qualified by the Corporation for assistance. Although our station claims more than 300 paid man-hours weekly, we have only one full-time staff member according to CPB's definition. (Although HEW recognizes the manufacturer's claim that our automation system is equivalent to two full-time staff members, CPB does not).

Therefore, we are not able to provide to our listeners the programming of the National Public Radio network; we are limited in the amount of financial support, if any, we may seek from CPB to increase and improve our service.

Our twelve colleagues in that four-state area are in exactly the same position—and always will be. They share our concern about the disparity in Corporation regulations which make no distinction between Los Angeles, California and Laramie, Wyoming. They share our concern about rigid regulations that seemingly ignore the special and peculiar needs of the small market audience.

We do not censure the Corporation for its efforts to improve public broadcasting in metropolitan markets. Yet, by imposing blanket rules, and having those rules apply with equal force in New York City and Missoula, Montana, the small market broadcaster is confronted—with his limited resources and limited staff—with conditions under which he simply cannot approach the goal of the Public Broadcasting Act to "make noncommercial educational radio and television services available to all citizens of the United States."

The anomaly here is that the small market public broadcaster is swimming upstream against a rising tide of regulations which allow a station in Chicago to receive a $25,000 grant while a station in the "boondocks" struggles to serve its audience—an audience which, it may be argued, has a greater need for public radio's services than its metropolitan counterpart.

Wilbur Schramm and Lyle Nelson, reporting for the Aspen Institute Program on "The Financing of Public Television," ask that we strike a "reasonable balance between local, regional and national support." That, too, is what we ask for public radio.

The Corporation must, in its long-range financing plans, allow for *substantial* and *continuing* financial support of small market stations.

The Corporation must recognize that many small market public stations can never meet the staffing criteria on their own. Collectively they must rely on massive contributions to reach the achievements demanded by CPB—and CPB itself must be the source for those funds.

The Corporation must recognize that need. The Corporation must include the small market broadcaster in its long-range financing plans in ways that it has not done in the past.

Schramm and Nelson concluded their study by noting that "the public interest requires a reasoned, dispassionate and farsighted approach." On behalf of the small market public radio broadcaster in the Intermountain West, I could ask no more of CPB.

Thank you.

Appendix G:
Emerging Technologies

The following material served as partial background for the chapter on "The Future—Alternatives and Recommendations" in the body of the text. It is included here in greater detail because of the significant potential effect of technology on the transmission of educational broadcast programming

Communication Transmission Systems

A variety of new transmission systems are now being developed. Some present a significant potential to supplement or even replace broadcast transmission. These include:

Specialized Communications Common Carriers

Cable Communications

Instructional Television Fixed Service

Satellites

Fiber Optics

Millimeter Wave Guides

Lasers

Communication Common Carriers

Bandwidth requirements depend upon the amount of information that must be transmitted in a given time. Thus, high-fidelity music broadcasts require 15 KHz, almost four times the 4-KHz bandwidth of telephone lines. Television, in turn, requires 4-1/2 MHz, over a thousand times the capacity of telephone lines.

The common carriers, therefore, provide the variety of services to accommodate both narrow and broadband requirements. AT&T, Western Union, General Telephone and Electronics, and some 1700 other telephone companies provide such services. These companies provide both narrowband and wideband facilities from teletype circuits, using 1/10 of a single voice grade channel, to full 6-MHz video channels of the type used for the PBS Interconnect and some of the regional and state distribution networks.

Narrowband Communication Channels. The most common narrowband facilities are telephone lines, used both for voice transmission and for data communications. Most educational facilities, of course, have direct access to the Direct Distance Dialing telephone network and use this service for transmission of administrative information. Some schools use network services to provide two-way voice interconnections with home-ridden students. Other schools use the telephone network to provide computer time sharing service both as an administrative and teaching tool.

Extensive utilization of the telephone network for specialized service (such as time sharing) tends to be very expensive. Information flow is restricted by the narrow bandwidth. As specialized common carrier services become available and as CATV systems become more common, the cost of broadband transmission should decline enough to provide greater flexibility.

Specialized Communications Common Carriers. In June 1971, the FCC created a category of carrier referred to as specialized communications common carriers. These services have evolved as an outgrowth of the expanded use of computers and data transmission from remote terminal locations. As early as 1958, the existing common carriers provided equipment to convert the digital output of a computer to a signal compatible with the switched telephone network. Demand for data-communication service increases with the number of computer installations and development of new computer applications.

Data transmitted on the existing common carrier facilities (an analog transmission network) are subject to numerous "errors" introduced along the transmission path. Errors are cumulative and the average error rate experienced on the direct distance dialing network is about one error in every 10^5 bits transmitted. While this small error is of little consequence to analog material, its presence in digital manipulations can be critical.

Digital transmission techniques allow for regeneration of the signal in a manner that does not reproduce noise introduced along the route. As a result, error in digital transmission is reduced to a level several orders of magnitude less than in analog transmission. For this reason, digital transmission is preferred for data communication.

Growth of Demand for Specialized Transmission Service. The demand for data transmission services became so great in 1963 that Microwave Communication, Inc. (MCI) petitioned the FCC for permission to construct a microwave link between Chicago and St. Louis to provide a variety of subscriber services. It was not until August 1969, however, that the FCC granted the construction permit for this route and then over strenuous objections from the established common carriers.

Between the time the FCC granted MCI its construction permit and July 15, 1970, when the FCC issued a Notice of Inquiry related to Specialized Common Carrier Service, there were thirty-seven separate proposals filed.[1] Of the thirty-seven, twelve were from MCI affiliates. These applications, many for networks serving the same areas, proposed over 1700 relay points.

The rationale for these filings was that specialized carriers could provide more diverse service, at lower rates, than could be provided by the existing carriers. The existing carriers, on the other hand, claimed that they were providing adequate service, and that introduction of such new service constituted duplication of service, inefficient use of resources, and siphoning of profits.

FCC Approval of Specialized Common Carrier Concept.   After about a year of evaluation and study by the FCC common carrier staff (June 1971), the commission agreed that more data transmission capacity and diversity of service was desirable, and that the potential of the specialized carriers should be allowed to develop. It was careful to state that the existing carriers and the specialized carriers would have to operate on a competitive basis and no group would be given a protected position.

Interest in entering this new area of communications was so great that by the end of February 1973 there were fifty-six proposals filed with the FCC, of which eighteen were approved and construction permits granted. Of the eighteen granted, MCI was given eight. It has had its Chicago-to-St. Louis route in operation for about two years. According to the MCI 1973 Annual Report, by the end of 1973 the company expects over 6 million channel miles of its microwave network to be installed, linking nineteen cities from New York to Los Angeles.

Services of the Specialized Carriers.   The services provided or planned by the specialized carriers are essentially point-to-point facilities providing a variety of bandwidths. Local loop arrangements and connection from terminal to carrier (figure G-1) must be made with existing carriers.

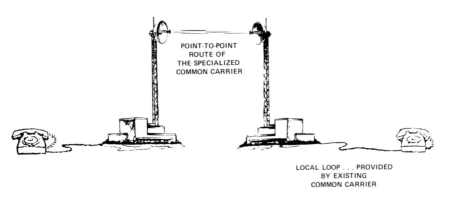

POINT-TO-POINT
ROUTE OF
THE SPECIALIZED
COMMON CARRIER

LOCAL LOOP . . . PROVIDED
BY EXISTING
COMMON CARRIER

Figure G-1. Service Interconnections.

The success of such specialized common carriers depends in large measure on the local services already provided by the existing common carrier. It should also be kept in mind that the entire concept of specialized carriers came about because of user dissatisfaction with the lack of alternatives afforded, at that time, by the existing carriers. The state of the art in data communications now provides greater transmission alternatives to the user, but these new alternatives are provided primarily by the existing carriers.

Existing Common Carrier Facilities.   The Bell System did not stand idle as these proceedings were in progress. In addition to its presentations before the FCC, it continued to develop plans for a digital data-transmission system to be in partial operation by mid-1974.

In October 1972, AT&T requested approval of a new digital network as the first phase of a nationwide digital data service (DDS). This first leg, given approval by the FCC early in July 1973, will link five cities: Boston, New York, Philadelphia, Washington, D.C., and Chicago. Initial service could be in operation by early 1974. The expanded service is expected to interconnect ninety-six cities by 1976.

DDS would provide digital channels for data rates of 2.4, 4.8, 9.6, and 56.0 kilobits/second. The initial system will provide synchronous transmission, whereas the expanded system will provide nonsynchronous service.

The initial leg of the DDS will also utilize a new Bell System developed transmission technique called Data Under Voice (DUV). DUV provides for transmission of 1.5 megabits/second over the generally unused lower portion of the existing microwave radio frequency spectrum. It is estimated from trial experience that even in microwave fade conditions, the DUV error rate will be less than one error in 10^7 bits (bps) sent. (On Bell System microwave routes, protection channels are provided in case weather or equipment malfunction causes an outage on the main microwave path. If a "fade" occurs, the transmission path is automatically switched to the protection channel.) Under nonfade conditions, transmission is expected to be virtually error free. The DUV technique will be utilized on both the microwave facilities and the coaxial cable carrier facilities of the Bell System.

As a result of hearings, inquiries, opinions, comments, the FCC has opened the way for competition with the existing common carriers; and the Bell System has developed a more extensive network design and greater diversity of data communication services. These factors, in total, have provided or promise to provide alternative transmission services to communications users not available several years ago.

It remains to be seen whether an open competitive atmosphere will develop in this area of specialized common carriers as a result of the FCC's action. In any event, it appears that the data-user will have a greater variety of transmission service alternatives available to him.

Multipoint Distribution Service (MDS). MDS is another type of microwave common carrier service, which was first established by the FCC in 1962. When instituted, MDS service provided a 3.5 MHz bandwidth, inadequate for video transmission. In 1971, the FCC removed the 3.5 MHz bandwidth restriction and made available the spectrum space between 2150 and 2160 MHz for video and data communication carriage.

An MDS system consists of a transmitter, studio facilities, a four-foot parabolic receiving antenna, and a "down-converter" for translating the super high frequency (SHF) to the VHF range (Channel 7 or 8). As common carriers, MDS systems operators have no control over the programming that they distribute over their system. The system is capable of transmitting video signals from cameras, film chains, or videotape in addition to being able to distribute signals from remote pickups. The range is limited to line-of-sight distances and by FCC ruling systems must operate at least fifty miles apart to prevent interchannel interference.

Cable Communication

Cable communication offers one of the best technical alterantives for the distribution of instructional, cultural, and public affairs information. At the same time it presents a flexibility that challenges the potential of programming and application diversity. The cable communications industry is gradually evolving from an unspectacular genesis as an extension of over-the-air television signals. The essential parts of a cable system (figure G-2) are its:

1. antenna to pick up over-the-air broadcast TV and radio signals,
2. headend facilities to amplify and channel the signals,
3. studios for live presentations, recording or the introduction onto the cable of film or videotape,

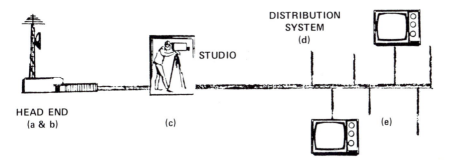

Figure G-2. Components of a Cable Television System.

4. distribution network of coaxial cable from the headend to the subscribers, and
5. subscribers' receiving equipment, which may incorporate a variety of devices for selective reception, pay TV, etc.

The details of CATV have been exposed in over five thousand documents, articles, and books. Many, in their enthusiasm, can best be compared to science fiction. At least six national newsletters distribute CATV information to a variety of audiences almost as diffuse as the audiences the "new" cable industry promises to reach. At the time of this research the industry had not yet achieved a separate Standard Industrial Classification code (it shares a category, Communication Services, Not Elsewhere Classified, within the major heading Communications, SIC 4899), but as industry revenues and equipment sales climb past the $450 million mark headed toward predicted levels of over $1 billion by 1980, it is reasonable to expect that the industry will be granted that mark of maturity.

Status of the Operating Cable Systems

As of August 13, 1973, there were 3033 systems providing services to 6041 communities.[2] In addition, 1662 franchises have been awarded but are not yet complete, and 2245 certificates are awaiting approval. In a recent survey completed in April 1973, the FCC indicates that cable television reaches 6,084,834 subscriber homes, or about 10 percent of the TV homes. Figure G-3 shows the growth in the number of households that subscribe to cable television service in the United States.

Cable System Ownership. The distribution of cable companies by category of ownership as of March 2, 1972 is shown in table G-1.

It should be noted that broadcasters are prohibited from owning cable

Table G-1
Cable System Ownership

Type of Ownership	Percentage of Total Systems
Commercial Broadcasters (Radio and TV)	37.9
Newspapers	6.3
Publishers	2.6
Film Producers	7.6
Theaters	3.4
Telephone Companies	2.0
Community or Subscriber Owned	2.9
Private Ownership	37.3
	100.0

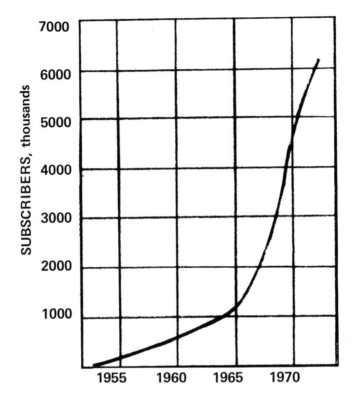

Figure G-3. Subscriber Growth Since 1952.

systems in the areas in which they are licensed to transmit an over-the-air TV signal, though overlap with radio signals is permitted. Since 1969 the percentage of cable TV systems owned by broadcasters has risen from 32 to almost 38. Traditionally, commercial broadcasters have vigorously resisted the growth of the cable industry. They see cable television as both an extension and a competitor to their medium, and they appear to be exerting every effort to gain as much control of the cable industry as possible. This is not an unfamiliar pattern. When broadcast TV became economically viable, it was the owners of broadcast radio systems who were both innovators and investors.

Like telephone service, cable communications is a type of "natural" monopoly—it is both difficult and wasteful to have two companies provide the same service to the same people. Such a condition tends to foster agglomeration and make the aggressively big companies get bigger. This is occurring in the cable industry. Of the 3033 systems in operation, the largest ten companies serve over 45 percent of the subscribers, or about 2.7 million homes. The top twenty-five serve over 60 percent of the subscribers, and the top fifty serve over 75 percent of the homes.

We determined from our site visits and telephone interviews that the opinions of the educational broadcasters range from the traditional commercial broadcasters' view that all CATV operators are "pirates," to one of recognizing cable's importance in extending their signals. Many educational broadcasters encourage cable systems to carry the ETV signals. Cable systems often carry the signals of the local ETV station, whether or not there is specific interaction between the educational station and the CATV system, because they recognize it as a selling point for their service.

Evolutionary Changes Within the Cable Industry. A number of changes within the cable industry in the past three to four years further indicate its process of maturing as an industry and presage the influence it is likely to have on the entire field of telecommunications. A brief exploration of these changes can help to put into perspective the effects of the growth of cable communications.

The Climate of Regulation. It is apparent, from the concern for regulation of the cable inudstry, that numerous agencies, special interest groups, and individuals are aware of the growth pattern and potential influence of the new medium. The most recent significant FCC rule (3rd Report and Order, March 1972) relaxed restrictions on the importation of commercial broadcast TV signals. The same order requires all new cable systems to provide minimum capability of twenty video channels. (Existing CATV systems have until 1977 to upgrade their channel capacity to that level.) In addition, until 1977 new systems are required to provide, free of charge on an experimental basis, one channel each for educational, public access, and municipal use.

The cable industry has complained that the FCC has not moved fast enough or far enough to let the cable industry grow. They insist that FCC policy appears to protect the commercial broadcast industry to the detriment of the cable industry.

State Regulation. In a recent burst of activity, a number of states have recognized the potential of cable TV and moved to regulate the industry. Such action ranges from total control of the franchise process in some states to mere advisory services in others. As of June 1973, seven states (Alaska, Connecticut, Hawaii, Nevada, New Jersey, Rhode Island, and Vermont) regulate the cable systems within their boundaries through public utility or public services commissions. In three others, Massachusetts, Minnesota, and New York, regulation is accomplished by a cable commission. In eleven other states (Alabama, California, Illinois, Maine, Maryland, Michigan, Oregon, Pennsylvania, South Carolina, Texas, and Wisconsin), study groups either have been formed or action is pending. Twenty-seven others have expressed various degrees of interest in cable regulation.

Local Regulation.   The cable franchise, with few exceptions, is granted
by the municipality that will receive the CATV service. As city councils acquire
sophistication, this franchising process is being performed much more carefully.
Competition for franchises in the larger cities is increasing, and citizens groups
are increasingly involved in shaping the franchise agreement. Special advisory
groups, such as the Cable Television Information Center, are available to provide
advice and help insure that the franchising process is responsible to the public
interest.

Technical Changes.   Several major technological advances are developing.
Perhaps the most important is the potential of two-way systems. This could
result in truly interactive cable communications, a much hoped for dream. In a
typical interactive system three major elements are envisioned: the computer,
the cable headend, and the subscriber terminal (figure G–4)

Within the home terminal is a keyboard through which the user communi-
cates with the system. The terminal may include a memory bank that maintains
the display. The most available and least expensive terminal is the touch-tele-
phone, which is already being applied to such areas as personal accounting and
record keeping. A television "frame grabber" could be incorporated, to select
specified frames of data from a stream containing thousands of discrete pictures.
As the picture is "grabbed" and displayed, the viewer is left with the impression
that his is the only picture transmitted.

Cable system interconnection is not far advanced, but interest is expanding
rapidly. The current state of cable technology provides a range of about twenty
miles from the headend. Some means is necessary to pass program material
from one such system to another. At present, a few contiguous systems are

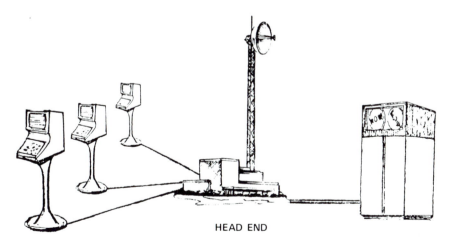

HEAD END

Figure G–4. Interactive System.

being interconnected by microwave facilities. Satellites might be used to receive program material transmitted from one point (well beyond the twenty-mile range) and beam the material to distant systems.

Several major system operators interested in developing satellite interconnection have assembled a task force to solicit industry cooperation. However, the FCC must further refine domestic satellite policy before such a system can be considered in any detail.

Service Changes. The most important new service prospect is "subscription television." A number of systems are being developed through which special programming is provided to the subscriber. Depending upon the particular system, the subscriber pays on a per program or on a monthly basis. At present this type of service provides an added inducement for new subscribers to a cable system in areas where over-the-air TV signals are adequate. More important in the long run, it provides a potential new source of income and consequent program diversity.

Picture, for example, the potential for financing educational programs through such subscription. A major difficulty of educational broadcasting revolves about the enormous and continuing cost of producing programs that present serious material *and* match the high standards of commercial broadcasting. Subscription television offers one of the most attractive prospects for replacing the current "grants-economy" of educational broadcasting. It should be noted, however, that during the early part of August 1973, the commercial TV broadcasters, through the National Association of Broadcasters, have instituted a significant renewal of their program to stop the growth of subscription television. They argue that the subscriber must pay for what he is now getting free on the commercial broadcast facilities. It is quite apparent that the broadcasters are reacting to the threat cable presents to their interests.

Other potentially significant new cable services are now being tested, including such things as retail purchasing from the home, alarm transmission, and health care services. In the health care delivery area alone, at least eight demonstrations are being conducted with the support of HEW. They range from the provision of medical advice to actual diagnosis and transmission of data such as pulse rate, heart beat, etc.

Satellites

The use of satellites as orbital relay stations has become an accepted reality. The elements of such a system are the ground station transmitter, the satellite itself, and the ground station receiver. The originating transmitter beams a microwave signal (uplink) of one frequency to the satellite receiver. The satellite changes the frequency and directs it to ground station receivers (downlink), thus providing a long-range distribution facility capable of transcontinental or transoceanic transmission.

The principal advantage of such transmission is the ability to supplement land-line facilities (cable or microwave) or to provide service in remote areas where it is economically impractical to run more conventional transmission facilities. Satellite transmission can also be received by a larger number of ground stations; however, they must be specially equipped. The major disadvantage of this technique is the short useful life of current satellites compared to the twenty-year-plus life expectancy of conventional telecommunication facilities.

Despite this, an experimental satellite will soon be launched to test its use in remote area education. The project, the sixth in a series of Applications Technology Satellites called ATS-F, is a combined effort of NASA, HEW's Office of Telecommunication Policy, the Corporation for Public Broadcasting, the National Institute of Education, and the Lister Hill National Center for Biomedical Communication. All of these agencies are working with the Rocky Mountain Federation of States, the Rocky Mountain Corporation for Public Broadcasting, the Appalachian Regional Commission, and the Alaska Office of Telecommunications.

The objective is to place, in synchronous orbit, a ton-and-a-half satellite that measures 52.5 feet between the tips of its solar energy panels. ATS-F will direct two TV channels in the 2500 MHz range at an area, referred to as a "footprint," 1000 miles long and 300 miles wide. In addition, each video channel will have four associated audio channels. The overall capability of the satellite will be as a transceiver for a full range of communications services: video, facsimile, radio, telephone, telegraph, and computer data.

The operations plan includes providing career-planning programming to some three hundred receiving antennas in the Rocky Mountain area, for direct reception or for distribution by other means, e.g., educational TV or CATV. Other programming will be supplied to the Appalachian region and to Alaska. Later, ATS-F will be moved over India for similar experiments.

Several International Telecommunications Satellite Consortium (INTELSAT) series satellites which connect the continents have been launched since 1965. The USSR orbited six communications satellites during 1972 to achieve the world's largest "domestic" satellite program.[3] Canada, with the launching of ANIK I in November 1972 and ANIK II in April 1973, becomes the second nation to have an active domestic satellite program.

The United States has lagged behind the other nations because standard operating procedures for the anticipated participants have been lacking. A number of communications companies, working in groups or alone, have expressed their interest by submitting proposals for systems, projecting launchings in 1974 or later. These include:

1.   AT&T and Communication Satellite Corporation (Comsat);

2. Western Union;
3. American Satellite Corporation and Western Union International;
4. RCA;
5. Comsat, MCI Communications Corporation, and Lockheed; and
6. Hughes and General Telephone & Electronics (GTE).

Utilization of Satellites for Interconnection of Telecommunication Services. Another promising use of the satellite will be to interconnect local telephone networks and cable television systems. Current cable systems have an inherent range limitation of approximately twenty miles, so that either conventional microwave systems or satellites are needed to interconnect them. A potential application is shown in figure G-5.

Significance to the Dissemination of Educational Programming. Video relay via satellite has been available for almost a decade. At the same time, other satellites survey cloud cover, act as storm surveillance units, and support a great variety of scientific experiments. Thus, satellites are a reality, and the feasibility of a "great antenna in the sky" cannot be denied. But before this potential becomes a *practical* reality for educational program dissemination, massive quantities of money and technical effort must be expended—and a real need must be demonstrated.

Fiber Optics

Fiber optics, sometimes referred to as "light pipes," are an intriguing and promising alternative to current electronic and radio frequency transmission. In this system a beam of light would be modulated to carry the television signal and then "piped" through a glass fiber a small fraction of the diameter of a hair. A small bundle of such fibers could contain thousands of separate light pipes, each carrying its own video signal.

Fiber optics is based upon the principle of total internal reflection of light along the axis of a filament of glass. A core of glass only a few microns in diameter is clad with a coating of a second glass with a slightly lower index of refraction than the core. The interface of the core and the cladding causes light directed down the filter to reflect back and forth (figure G-6) in accordance with Snell's principle of refraction: A light ray aimed from a medium with a high index of refraction will be *totally reflected* if its angle of incidence is greater than a "critical angle."

The development of fiber optics for communications transmission is fairly recent. This medium is easily capable of transmitting one video channel per fiber. The first reasonably low-cost fibers were developed by Corning Glass in 1970. At that time researchers had developed a method of drawing glass fibers

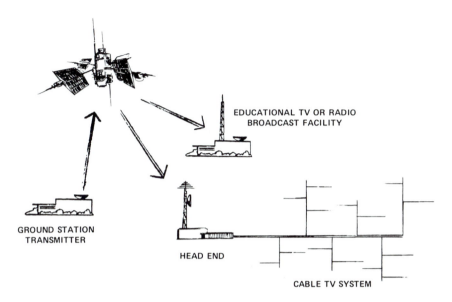

EDUCATIONAL TV OR RADIO
BROADCAST FACILITY

GROUND STATION
TRANSMITTER

HEAD END

CABLE TV SYSTEM

Figure G-5. Potential Application of Satellites for Interconnection.

with loss characteristics equal to about 20 db/km. More recent research efforts by Corning and Bell Telephone Laboratories (BTL) have resulted in clad fibers with losses as low as 2 db/km. At the same time, BTL has developed a single-material fiber system (figure G-7) with losses of 5 db/km, consisting of a hollow tube, a solid inner core, and a support plate for the core, all of the same low-loss glass. These losses are roughly comparable to the attenuation of other transmission media. According to BTL, the tensile strength of a single fiber when first drawn ranges up to several million pounds per square inch, or greater than steel. Thus, fibers in bundles should have excellent strength and durability.

CLADDING

CORE

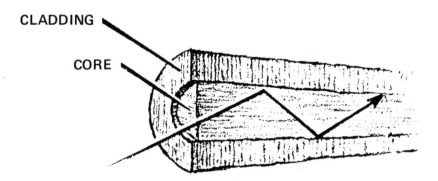

Figure G-6. Reflected Light in Optical Fiber.

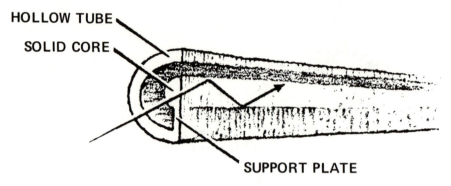

HOLLOW TUBE

SOLID CORE

SUPPORT PLATE

Figure G-7. Single-Material Optical Fiber.

Assuming a continuation of the refinement of the fiber optic techniques and development of the necessary modulators, we can expect initial commercial transmission utilization by 1980.

At the same time, techniques for modulating laser beams are also being developed. Multiplexed laser beams could be transmitted by the fiber, and repeaters used to regenerate the signal back to its original form every few miles, thus providing a high-capacity, low-loss transmission path with almost unlimited bandwidth. Even in the telephone network, fiber optic transmission appears to be a likely substitute for copper wire.

Millimeter Waveguide

In another search for broadband transmission techniques, BTL is developing millimeter wave transmission. This *extremely* high-frequency radiation (from 40 to 110 GHz) is modulated by pulse coding techniques and directed down a circular waveguide. Millimeter wave transmission does not require the frequent (about every two miles) amplification of wire line or optical systems, but uses a regenerative repeater about every twenty miles. Each such waveguide can carry the equivalent of more than a hundred video channels. Because the geometry of the waveguide is critical, it must be buried in a protective conduit. While the millimeter waveguide technique requires special methods of installation, it does appear to be a possible alternative method of transmission.

Lasers

Laser technology appears as a component in a number of communications devices such as facsimile, video recording devices, and optical character recognition equipment. However, research and development on open laser beams as a mode of transmission has not been too encouraging. Open systems are subject to interference and noise and are not as efficient as other methods of optical transmissions. (See section on fiber optics.)

Recording Mechanisms

Video Recording. The Sony video cassette has received rapid acceptance in many market areas. We find, as a result of other research we are conducting, that the video cassette recorder/player is currently being used extensively in instructional settings. The recorder allows the teacher great freedom in the time and presentation format of video material. In some cases recorders are being used to take the material off the air for playback at the teacher's convenience. Without this, the teacher must accommodate his class activity to the schedule of the local educational broadcasting station or cable system. Industrial use of the tape player/recorder for training appears to be growing at an equally accelerated rate.

The large home-user market anticipated (or hoped for!) by the video record industry has not yet materialized and does not seem likely in the near future. The required conditions we noted in earlier reports [4] still have not occurred: truly low-cost program material and hardware in the "under $500 range." The "softnesss" of the home video player market is evident in the announcement on July 9, 1973, that AVCO Corporation had filed bankruptcy proceedings for its Cartridge Television subsidiary. Cartridge TV's product was the Cartravision home video player system. AVCO cited lack of available funds (inadequate sales) and continued losses as the reason.

At present a number of cassette and cartridge videotape recorder/players (and a number of tape sizes) are available. For example:

Cassettes

1 in.: International Video Corp. (IVC)

3/4 in.: Sony, Panasonic, and Japan Victor Co. (JVC)

1/2 in.: Sanyo (announced but not commercially available)

(Ampex and RCA provide a professional 2 in. cassette VTR)

Cartridges

1/2 in.: Panasonic, Shibaden, Concord, and Norelco. (Norelco is not EIAJ1 standard as are the other three.)

In the fall of 1972, Matsushita Electric announced a "video" cassette system that provides up to 1000 still color pictures plus sound on a standard Philips 60-minute cassette (1/8 in.) tape. The system does not use helical scan techniques but rather a new horizontal scan. This is a laboratory device and to date no marketing plans have been announced. Another device using low tape speed and sophisticated encoding techniques is said to be under development by Arvin Corporation.

Table G-2 provides basic data on the major magnetic videotape devices.

Videodisc Systems.   Several videodisc systems have been introduced, but we have great doubts about the cost and reliability of these systems based on their requirement for highly precise mechanical components. The introductory price of one player, for example, was estimated at about $150. Within a brief time this introductory price was raised to over $1000.

Teledec.   Telefunken and Decca, Ltd. have developed a system called Teledec, which records color by separating the high-frequency information from the low-frequency portion of the video signal and utilizes the latter portion as a line-sequential color signal. The decoding system can be designed to match any color signal so that the same videodiscs can be used on equipment designed for the particular country (subject to the varying frame rate).

The sound record is carried entirely within the blanking period on the disc. All of the information is carried on a "hill and dale" track of a disc spinning at 1800 rpm. This track is detected by a pressure transducer to bypass the problems of stylus inertia and record wear.

The playing time is five minutes for a nine-inch disc and twelve minutes for a twelve-inch disc. Teledec proposes to solve the limited playing time by an "instant" (less than one-second) change mechanism. The discs are contained in a cartridge inserted into the machine. The mechanism will withdraw the in-

Table G-2
Videotape Recorders–Helical Scan Devices

Manufacturer	Tape Width in.	Maximum Playing Time, min.	Tape Format	Price, $
AKAI	1/4	80	Open Reel	940–1700
Cartridge TV	1/2	120	Cartridge	1700 (max)
Panasonic	1/2	30	3/4 in. Cassette	950–1375
	3/4	60		
Philips	1/2	40	Cassette	1295
RCA	3/4	60	Cassette	700 (est)
Hitachi	1/2	60	Cassette	775–1650
Sony	1/2	30	Reel-to-reel	995–1395
	3/4	60	Cassette	
Victor (JVC)	1/2	30	Reel-to-reel	1400 (max)
	3/4	60	Cassette	

Source: Derived from *Electronics* (4 January 1973), p. 47.

dividual records from the magazine, loading each one while the previous disc is playing. A price of $1100 has been quoted.

The problems of establishing production and maintaining quality control must be formidable. Track separation is 0.008 mm (approximately one-tenth the diameter of a human hair), so concentricity, distortion of the disc, and uniformity and stability of the mechanical drive must be maintained at incredibly close tolerances. Any error in tracking or pickup wear will result in jumping or mixing of consecutive frames. The quoted record production cost (under $3/hour) is an order of magnitude below other systems.

RCA LP Videodisc. RCA has announced a "low cost" videodisc system, but to date has not presented it in formal public display. In informal conversation, RCA indicates a possible introduction of production units in 1975.

The RCA system uses a "needle-in-groove" on a rigid disc that can be recorded on both sides. Each side contains up to twenty minutes of video and stereo sound. The disc revolves in the 400-to-500-rpm range. The price of the player is expected to be in the $150 to $400 range. According to a report in *Television Digest* the critical process is the making of the master videodiscs.[5] The replication of the disc is conventional, and costs of the copies are said to be about $0.05 each.

Philips VLP Disc (Video Long Play). The Philips VLP disc system has been publicly demonstrated since mid-1972. It employs a disc that can be recorded on both sides, with a playing time per side of up to about forty-five minutes. Video and audio information is contained in microscopic indentions, the length and spacing of which determine chrominance, luminance, synchronization, and sound. The indentions are scanned by a low-cost 1-mw helium-neon laser. The disc base is vinyl with an aluminum coating. The player is capable of stop frame, slow motion, reverse, and fast forward and can also be indexed for random access.

The sytem is made to accommodate both the European PAL color standard and the U.S. NTSC color system. The U.S. version revolves the disc at 1800 rpm to accommodate the thirty-frame U.S. television standard. The estimated cost of the player is $625.

The Philips player also appears to require a mechanically complex device to "read," within allowable tolerances, the microscopic information indentions. Video reproduction is adversely affected by disc warping, necessitating additional complex mechanical compensation.

MCA Disco-Vision. MCA has demonstrated a disc system with a playing time of about twenty minutes per side. The demonstration showed freeze frame and fast forward and reverse functions. Other than information that the scanning of the disc is accomplished by a laser and the groove tolerances are very high, almost no system details are available. No information has been made public as to when units will be available.

Arvin Video Discassette.   The Arvin system uses a disc encased in a plastic housing. Information is recorded onto the reusable magnetic coating of the disc, up to a maximum of 300 still frame images. The playback can be at one, three, six, ten, and fifteen frames per second.

Production units of the Arvin system are available for $4000, with the Discassettes selling for $25 each. There are several commercial applications, such as display of hospital patient records, in which this system is used.

Other Disc Systems in Development.   There appears to be a flurry of activity in videodisc development. Another innovative approach, which is still in the research stage, was announced by Sydnor-Batent-Scanner in Albuquerque. This system utilizes numerous minute lenses and photographic images. A single disc is said to have the capacity of ninety minutes of black-and-white images or forty-five minutes of color.

Conclusions Concerning Videodiscs.   We believe that there may be a commercial application for the disc approach, but that there is less likelihood of a residential market in the near future than for the magnetic tape system. So far, none of the discs are "home recordable" as is possible (if not yet economical) with the magnetic tape devices. In addition, the criteria mentioned earlier (i.e., low-cost program material and hardware costing less than $500) probably will not be met for some time, perhaps five to ten years.

Another factor in the market demand for videodisc systems centers about the question: "How many times will I use a video recording?" A recent proprietary study concluded that entertainment material will be used less than twice on the average. This casts considerable doubt on the probable market for entertainment videodiscs. In our opinion, current systems will be a curiosity for some time and may find some special applications that are not obvious now. But for the next ten years, videotape will be used more extensively.

Microimaging.   Microimaging has been receiving great attention from equipment developers and marketing groups. No significant breakthrough has been made recently, however, that would open up the area to new applications. This technique of storage is fairly static and, as such, does not represent a new area of technology that will have a foreseeable impact on educational telecommunication.

Non-Impact Printing.   Non-impact printing comprises two printing techniques: electrostatic and thermal. Electrostatic techniques include the dry toner processes and the liquid ink-jet method. In the first, used in copy duplication, a reflected image from an original document causes a corresponding portion of a piece of plain paper to hold an electrostatic charge. Toner powder is made to flow over the paper and is attracted by the charge while surplus toner is removed. The paper with the adhering toner is "fixed" to make the toner permanent.

In the ink-jet process, extremely small droplets of liquid ink are forced out of a nozzle arrangement. An electrostatic charge is applied to the droplet and it is drawn toward a paper substrate by a difference in electrical potential. Control grids between the nozzle and the substrate deflect the droplet. If a character is to be printed, the grids will place it in one position of a dot matrix; successive droplets will complete the character. The importance of the ink-jet printing technique is that the character generation is controlled by digital signals. Thus, it provides a good possibility for terminal print-out devices in interactive telecommunications networks.

Terminal Devices. Interactive communication is a much sought after goal, but the only extensive experience is through telephone lines. It is not yet clear how broadband capability may prove useful, if at all. Potential applications depend upon the characteristics and availability of appropriate terminal equipment and switching hardware through which to center and channel the information. And this in turn waits upon the definition of viable applications. A variety of new terminal devices are being developed to accommodate both institutional and home use.

Videotape Terminals. A significant number of school systems have acquired or plan to purchase videocassette equipment. Such equipment is being used to record off-the-air transmissions of instructional material for flexibility in schedule and curriculum. This provides the school a rather inexpensive capability to build a tape library.

Videocassette devices are available in both 1/2-inch and 3/4-inch helical scan tape formats and range in price from about $800 for a player only to about $1400 for a recorder/player. One company provides a player/recorder with a timing device for automatic recording.

Interactive Cable Terminals. The cable communication industry is developing a variety of terminals necessary for residential two-way services. Three companies have announced plans to market terminal equipment in the near future: Jerrold, Theta-Com, and EIE (a Division of RCA). A typical terminal looks like figure G-8. More complex (and more expensive) units might include a channel selection switch, an alpha-numeric keyboard, and a paper printout mechanism to provide full interactive capability.

It is generally held that fully interactive switched video interconnection is not economically feasible. Even Picture Phone[a] has proven too expensive for residential use, and it is not certain that a picture will enhance all types of com-

---

[a]Registered Service Mark of AT&T Co.

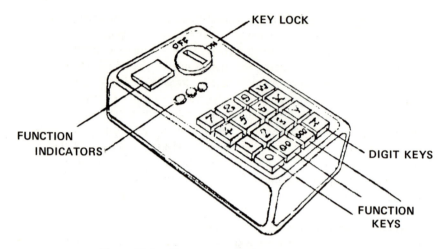

Figure G–8. Typical Small Interactive Terminal.

munication (some experiments are negative). In our opinion a more likely
system will combine video distribution with narrowband response—rather than
two-way video.

The broadband capacity of cable systems enhances the possibility of using
facsimile devices for the distribution of "hard copy," perhaps in obtaining
material from a library resource center. But again, the cost of facsimile transmission is too high to make this application practical for all but the most
specialized needs.

Component Devices.   The past two decades have seen spectacular advances
in the design and production of integrated electronic components. A new low-
cost ($300) calculator contains over 20,000 transistors! The vast potential of
such new components includes the possibility of new advances in low-light-level
video cameras, mass memory devices, and more versatile display systems. As
these devices leave the laboratory, we can expect to see smaller, more efficient
TV cameras; high-density, low-cost storage devices; and less expensive display
systems. One cannot discount the possibility of elaborate interactive systems
based on such development.

Notes

Chapter 2
The Goals of Educational Broadcasting

1. Amendment to Communications Act of 1934, Public Law 87–447, sections 390 and 392(d).

Chapter 3
The Status of Educational Broadcasting

1. The Educational Broadcasting Facilities Program is currently assigned to the National Center for Educational Technology, DCSS, U.S. Office of Education. Delegations have included: Office of Administration; Bureau of Elementary and Secondary Education; Bureau of Adult and Vocational Programs; Bureau of Adult, Vocational, and Library Programs; Bureau of Libraries and Educational Technology; D.C. Development; and D.C. School Systems.
2. As listed by channel in *Broadcasting Yearbook, 1973,* 8.6 percent of VHF stations and 20.8 percent of UHF stations are educational stations.
3. *Cable Television and Education—A Report from the Field* National Cable Television Association, Washington, D.C. (March 1973).
4. R.G. Noll, M.J. Peck, and J.J. McGowan, *Economic Aspects of Television Regulation,* The Brookings Institution, Washington, D.C. (1973).
5. W. Schramm and L. Nelson, *The Financing of Public Television,* Aspen Institute Program on Communications and Society, Palo Alto, Calif. (1972).
6. The basis for most of this material is N. Katzman, *One Week of Public Television, April 1972,* Corporation for Public Broadcasting, Washington, D.C. (1973).
7. Based on Dale Mark Rhodes, *The Utilization of Television Service in North Carolina's Public Schools,* The University of North Carolina Television Network, Chapel Hill (1972).
8. *The Viewing of Public Television–1971,* Prepared for The Corporation for Public Broadcasting, Louis Harris and Associates, Inc., study No. 2135 (November 1971).

Chapter 5
Summary

1. F.J. Doyle and D.Z. Goodwill, *An Exploration of the Future in Educational Technology,* Bell-Canada (1971).

Appendix G
Emerging Technologies

1. FCC 70–768 49876, Docket No. 18920 "Notice of Inquiry to Formulate Policy, Notice of Proposed Rule Making, and Order" (July 15, 1970).
2. Weekly CATV Addenda, 435, *Television Digest,* August 13, 1973.
3. Table of Artificial Satellites Launched in 1972, supplement, *Telecommunications Journal,* 40, no. 4 (April 1973).
4. G.W. Tressel and D.P. Buckelew, *A Discussion of the Video Recording Industry and Its Future,* Battelle-Columbus Laboratories, Communication Research Laboratory, Columbus (1971): and D.P. Buckelew and G.W. Tressel, *A Survey of New Developments in Broadband Communications and Video Recording,* Battelle-Columbus Laboratories, Communication Research Laboratory, Columbus (1973).
5. *Television Digest,* 12, no. 30 (24 July 1972): 7.

Bibliography

Books

Borchardt, K., *Structure and Performance of the U.S. Communication Industry.* Cambridge: Harvard University Press, 1970.

Chu, G.C., Schramm, W. *Learning from Television.* Washington, D.C.: National Association of Educational Broadcasters, 1968.

Noll, R.G., Peck, M.J., and McGowan, J.J. *Economic Aspects of Television Regulation.* Washington, D.C.: The Brookings Institution, 1973.

Smith, R.L. *The Wired Nation.* New York: Harper & Row, 1972.

Udell, G.G. (compiler) *Radio Laws of the United States.* 1972 Edition. Washington, D.C.: U.S. Government Printing Office, 1972.

Reports

Anderson, J.A. *Operations and Costs—A Study of Educational Public Television Stations.* Washington, D.C.: National Association of Educational Broadcasters, 1973.

———. *Public Television Station Projected Operations and Costs.* Washington, D.C.: National Association of Educational Broadcasters, 1973.

Badger, V.M. *The College Radio Station—Report on Student-Operated Radio Stations at Colleges and Universities.* Washington, D.C.: Corporation for Public Broadcasting, 1969.

Carnegie Commission on Educational Television. *Public Television—A Program for Action.* New York: Bantam Books, 1967.

Corporation for Public Broadcasting. *Public Radio Financial Statistics—Fiscal Year Ending June 30, 1972.* Washington, D.C., 1973.

———. *Financial Statistics of Public Television Licensees—Fiscal Year Ending June 30, 1972.* Washington, D.C., 1973.

Dickson, E.M., Bowers, R. *The Video Telephone—A New Era in Telecommunications.* Ithaca, N.Y.: Cornell University Press, 1973.

Doyle, F.J., Goodwill, D.Z. *An Exploration of the Future Educational Technology.* Bell-Canada, 1971.

FCC Cable TV Census—1972. Washington, D.C.: Federal Communications Commission, April 1973.

Ford Foundation. *Focusing on the Viewers Feedback* (WJCT), Jacksonville, Florida. New York, 1972.

Frymire, L.T. *Study of State Public TV Systems.* Washington, D.C.: Corporation for Public Broadcasting, 1969.

Katzman, N. *One Week of Public Television—1972.* Washington, D.C.: Corporation for Public Broadcasting, 1973.

Lee, S.Y., Pedone, R.J. *Broadcast and Production Statistics of Public Television Stations and Licensees: Fiscal Year 1971.* Washington, D.C.: Corporation for Public Broadcasting, 1972.

——. *Financial Statistics of Public Television Licensees.* Washington, D.C.: Corporation for Public Broadcasting, 1972.

——. *Summary Statistics of CPB-Qualified Public Radio Stations, Fiscal Year 1971.* Washington, D.C.: Corporation for Public Broadcasting.

Linville, W.K. *The Role of Telecommunications in the Regional Delivery of Education Services.* Washington, D.C.: Department of HEW, 1972.

Louis Harris and Associates, Inc. *The Viewing of Public Television—1971.* Washington, D.C.: Corporation for Public Broadcasting, 1971.

Long-Range Financing of Public Broadcasting. Washington, D.C.: Corporation for Public Broadcasting, March 1973.

McKay, B., *KQED and Its Audience.* Stanford, Calif.: Stanford University, 1971.

National Cable Television Association. *Cable Television & Education.* Washington, D.C., 1973.

O'Neill, J.J. *Testing the Applicability of Existing Telecommunications Technology in the Administration and Delivery of Social Services.* Washington, D.C.: Department of HEW, 1973.

PBS on Record—The Public Broadcasting Service Programming, October 1971-October 1972. Washington, D.C.

Pelnik, C. Glixon, H.R. *A Planning Document for the Establishment of a Nationwide Educational Telecommunications System.* Washington, D.C.: Synergetics Inc. for the Department of HEW, 1972.

Rhodes, D.M. *The Utilization of the In-School Television Service in North Carolina's Public Schools.* Chapel Hill: University of North Carolina, 1972.

Robertson, J. *The Future Role of Regionals.* Washington, D.C.: Corporation for Public Broadcasting, 1970.

Schramm, W. Nelson, L. *The Financing of Public Television.* Palo Alto, Calif.: Aspen Institute Program on Communications and Society, 1972.

Steiner, R.L. *Visions of Cablevision.* Cincinnati: Stephan H. Wilder Foundation, 1972.

United States Department of HEW, Joint Council on Educational Telecommunications. *Current Status and Future Possibilities of the Educational Broadcast Facilities Programs and Related Programs.* Washington, D.C., 1971.

——. *Educational Television—The Next Ten Years.* Washington, D.C., 1962.

Publications

Burke, J.E. "The Public Broadcasting Act of 1967," *Educational Broadcasting Review,* vol. 6, nos. 2, 3, and 4 (1972).

Hess, P.J. "Montana's ETV Quandary," *Montana Journalism Review*, no. 14 (1971).

Macy, J. "The Short and Unhappy Life of P.B.S." *The Center Magazine* 6, no. 3 (May/June 1973): 52, 53.

Millard, S. "A Special Report on Public Broadcasting," *Broadcasting* (8 November 1971).

Broadcasting Yearbook 1973, Broadcasting Publications, Inc., Washington, D.C. (1973).

NAEB Telecommunications Directory 1973, National Association of Educational Broadcasters, Washington, D.C.

Television Digest with Consumer Electronics, Television Digest, Inc., Washington, D.C. (1973); CATV Addenda; Television Factbook Edition No. 42, 1972-1973; Preproduction Material, Edition No. 43, 1973-1974; CATV & Station Coverage Atlas—1973.

Selected issues of:

Electronics

Communications News

Science

Bell Laboratories Record

CATV

TV Communications

Television Digest

Wall Street Journal

NCTA Bulletin

Telecommunications

Broadband Communication Report

Telecommunications Journal

FCC Information Bulletin

Telephone

IEEE Transactions on Communications

Miscellaneous

Communication Act of 1934.

Docket No. 16495—Before the FCC in the matter of: Establishment of domestic communications satellite facilities by non-governmental entities. Reply comments by The Ford Foundation in response to FCC Notices of Inquiry of March 2, 1966 and October 20, 1966.

Docket No. 18920, Notice of Inquiry to Formulate Policy, Notice of Proposed

Rulemaking, and Order, FCC, Washington, D.C., (1970) in the matter of
Specialized Communications Common Carriers.
Educational Television Broadcasting Act of 1962 (PL 87–447).
Public Broadcasting Act of 1967 (PL 90–129).
Title 47, US Code 390.
Hearings:
Before The Subcommittee on Communications of the Committee on
Commerce, U.S. Senate, 93rd Congress, First Session on S 1090 and S 1228.
28, 29, 30 March 1973.
Senate Report No. 93–123, Additional Views to Accompany S 1090–
Authorization for Public Broadcasting, 17 April 1973.
House of Representatives Report No. 93–324, Public Broadcasting to
Accompany H. 8538, 22 June 1973.

Index

About the Authors

George W. Tressel manages the Educational Development Section of the Center for Improved Education of Battelle's Columbus Laboratories. He directs the activities of Battelle's Communication Research Laboratory, which he established in 1967. Mr. Tressel was graduated in Physics from the University of Chicago in 1943, and since that time has conducted a wide variety of communication activities. He has produced numerous award-winning scientific and technical films as an independent producer, and at Battelle and Argonne National Laboratory. He has also headed a wide variety of communication policy studies for federal and state government agencies. Mr. Tressel is an Adjunct Professor at Ohio State University and an associate of the Academy for Contemporary Problems.

Donald P. Buckelew is director of the Municipal Assistance and Policy Development Division of the New York State Commission on Cable Television. An expert on communication technology and economics, Mr. Buckelew has conducted research on community antenna television (CATV), video record devices, educational television, telephone and peripheral devices, and other data-gathering or transmission systems. Mr. Buckelew is a graduate of Seton Hall University.

John T. Suchy is a research communications scientist with the Battelle Center for Improved Education. He is a graduate of the University of Montana and holds the M.A. degree from the University of Iowa. Under a Fulbright Scholarship, he conducted comparative studies of television programming in the United States and Great Britain. Mr. Suchy has written professional articles on communications and scientific subjects for a wide variety of publications and has written the scripts of some 40 television programs and technical films. He joined Battelle in 1970, coming from Argonne National Laboratory where he served on the staff of the Laboratory Director.

Patricia L. Brown is senior advisor for Information Systems Design in the Battelle Center for Improved Education. A graduate of the University of Southwestern Louisiana, she holds the M.A. degree from The University of Texas. Her

125

entire professional career has been in the field of information transfer. Miss Brown is a recognized expert in information system analysis, design, operations, and reporting, with more than 20 years' experience in technical libraries, information services, survey design, project management and reporting, and educational materials development.